BOSTON STUDIES IN THE PHILOSOPHY OF SCIENCE

VOLUME XXXVII

HERMANN VON HELMHOLTZ

EPISTEMOLOGICAL WRITINGS

SYNTHESE LIBRARY

MONOGRAPHS ON EPISTEMOLOGY,

LOGIC, METHODOLOGY, PHILOSOPHY OF SCIENCE,

SOCIOLOGY OF SCIENCE AND OF KNOWLEDGE,

AND ON THE MATHEMATICAL METHODS OF

SOCIAL AND BEHAVIORAL SCIENCES

Managing Editor:

JAAKKO HINTIKKA, *Academy of Finland and Stanford University*

Editors:

ROBERT S. COHEN, *Boston University*

DONALD DAVIDSON, *University of Chicago*

GABRIËL NUCHELMANS, *University of Leyden*

WESLEY C. SALMON, *University of Arizona*

VOLUME 79

Pastel by von Lenbach, 1894

HERMANN VON HELMHOLTZ (1821–1894)

(By Courtesy of Friedr. Vieweg & Sohn, 1977)

BOSTON STUDIES IN THE PHILOSOPHY OF SCIENCE

EDITED BY ROBERT S. COHEN AND MARX W. WARTOFSKY

VOLUME XXXVII

HERMANN VON HELMHOLTZ

EPISTEMOLOGICAL WRITINGS

THE PAUL HERTZ/MORITZ SCHLICK CENTENARY
EDITION OF 1921
WITH NOTES AND COMMENTARY BY THE EDITORS

NEWLY TRANSLATED BY MALCOLM F. LOWE

EDITED, WITH AN INTRODUCTION AND BIBLIOGRAPHY, BY

ROBERT S. COHEN AND YEHUDA ELKANA

D. REIDEL PUBLISHING COMPANY

DORDRECHT-HOLLAND / BOSTON-U.S.A.

Library of Congress Cataloging in Publication Data

Helmholtz, Hermann Ludwig Ferdinand von, 1821–1894.
    Epistemological writings.

    (Boston studies in the philosophy of science ; v. 37)
(Synthese library ; 79)
    Translation of Schriften zur Erkenntnistheorie.
    'Works by Helmholtz': p.
    Bibliography: p.
    Includes index.
    1. Knowledge, Theory of. I. Hertz, Paul, 1881–   II. Schlick,
Moritz, 1882–1936. III. Lowe, Malcolm F. IV. Cohen, Robert Sonné.
V. Elkana, Yehuda, 1934–   VI. Title. VII. Series.
B3279.H53 S3713   121   77-20909
ISBN 90-277-0290-X
ISBN 90-277-0582-8 pbk.

Translated from

SCHRIFTEN ZUR ERKENNTNISTHEORIE

first published by Julius Springer, Berlin, 1921

Published by D. Reidel Publishing Company,
P.O. Box 17, Dordrecht, Holland

Sold and distributed in the U.S.A., Canada and Mexico
by D. Reidel Publishing Company, Inc.
Lincoln Building, 160 Old Derby Street, Hingham,
Mass. 02043, U.S.A.

Printed in The Netherlands

TABLE OF CONTENTS

INTRODUCTION

[1977]

Hermann von Helmholtz in the History of Scientific Method

In 1921, the centenary of Helmholtz' birth, Paul Hertz, a physicist, and Moritz Schlick, a philosopher, published a selection of his papers and lectures on the philosophical foundations of the sciences, under the title *Schriften zur Erkenntnistheorie.* Combining qualities of respect and criticism that Helmholtz would have demanded, Hertz and Schlick scrupulously annotated the texts. Their edition of Helmholtz was of historical influence, comparable to the influence among contemporary mathematicians and philosophers of Hermann Weyl's annotated edition in 1919 of Riemann's great dissertation of 1854 on the foundations of geometry.

For several reasons, we are pleased to be able to bring this Schlick/ Hertz edition to the English-reading world: first, and primary, to honor the memory of Hermann von Helmholtz; second, as writings of historical value, to deepen the understanding of mathematics and the natural sciences, as well as of psychology and philosophy, in the 19th century – for Helmholtz must be comprehended within at least that wide a range; third, with Schlick, to understand the developing empiricist philosophy of science in the early 20th century; and fourth, to bring the contributions of Schlick, Hertz, and Helmholtz to methodological debate in our own time, a half century later, long after the rise and consolidation of logical empiricism, the explosion of physics since Planck and Einstein, and the development of psychology since Freud and Pavlov.

Helmholtz deserves renewed critical attention from scientists and philosophers. His masterful scientific achievements, even aside from his philosophical reflections upon concepts and methods, exhibit nearly all of the contrasting and contending methodological characteristics of the 19th century. Hertz and Schlick chose to focus their attention (and their Foreword speaks to this) upon a significant but limited portion of Helmholtz' interests, those in geometry, measurement, and perceptual cognition. Others have studied Helmholtz as a whole; we shall refer to

Meyerson, Cassirer, the Marxist philosophers Hörz, Wollgast, and Kosing, and others.

* * *

Like the physicists Heinrich Hertz and Enrico Fermi, Helmholtz was both theorist and experimentalist, but the breadth of his investigations was far greater, for he was a mathematician, physiologist, physicist, and psychologist. He worked in the cognitive and sensory psychology of hearing and vision, theoretical mechanics, electrodynamics, physical and chemical thermodynamics, electrochemistry, geometry, applied mathematics, medical experimentation and instrumentation, hydrodynamics, esthetics of musical theory and painting, atmospheric physics, epistemology, and more. When in need of conceptual clarification, Helmholtz wrote exemplary studies in the history of science, the most penetrating of these being his address of 1884, 'On the history of the discovery of the principle of least action'. His scientific writings were immense in number, and were influential throughout the 19th century. His two great treatises were, as Kelvin observed, each the 'Principia' of their field: in optics, the great *Physiological Optics* (3 vols; 1856, 1860, 1867), and in acoustics, physiology of hearing, and the theory of music, *Sensations of Tone* (1863). Helmholtz' research papers demonstrate mathematical skill, conceptual imagination, critical elucidation, profound inductive as well as deductive sensitivity, technical inventiveness, and a warm respect for the work of others. A selection of titles will suggest how extraordinary a scientist he was.[1]

1843, 'On the nature of fermentation and putrefaction' [his first publication, a pioneering step on the road to Pasteur]
1846, 'Physiological heat'
1847, 'On the conservation of force' [but 'force', *Kraft*, soon was taken to mean 'energy'; this is the classic memoir on conservation of energy]
1850, 'Rate of transmission of the excitatory process in nerves'
'On the methods of measuring very small intervals of time, and their application to physiological purposes' [These papers de-

monstrated what his teacher, the superb Johannes Müller, had thought impossible, the measurement of nerve processes]

1851, *Description of an Ophthalmoscope for the Investigation of the Retina in the Living Eye* [Helmholtz invented the fundamental diagnostic tool for medical investigation of the eye]

1852, 'On Brewster's new analysis of solar light'

1853, 'On the scientific researches of Goethe' [He went further in these investigations in 1892]

1854, 'On the interaction of natural forces'

1855, 'On the accommodation of the eye'

1856, 'On the explanation of lustre'

1858, 'On integrals of the hydrodynamic equations which express vortex-motions' [a classic analysis]

1859, 'On the quality of vowel sounds'

1861, 'On the application of the law of conservation of energy to organic nature'

1865, 'On the properties of ice'

1867, 'The relation of optics to painting' [a fundamental investigation]
'The mechanics of the auditory ossicles and of the membrane of the tympanum'

1868, 'On the origin and significance of the axioms of geometry' [in this volume]
'On the facts underlying geometry' [in this volume]

1869, 'The facts in perception' [in this volume]
'On hay fever' [he suffered somewhat and decided to investigate]

1870–71, 'On the theory of electrodynamics' (3 parts) [the major communication of Maxwell's theory to the continental physicists, but of original calibre]

1873, 'The theoretical limit of the efficiency of microscopes'

1875, 'Whirlwinds and thunderstorms'

1877, 'On thought in medicine' [another classic]
'On academic freedom in German universities'
'On galvanic currents caused by differences in concentration: deductions from the mechanical theory of heat'

1881, 'On the forces acting on the interior of magnetically or dielectrically polarized bodies'
'On the modern development of Faraday's conception of electric-

ity' [the first statement of the fundamental atomicity of electricity]

1882, 'The thermodynamics of chemical processes'

1884, 'Principles of the statics of monocyclic systems'

'On the physical significance of the principle of least action' [see below]

1887, 'Further researches concerning the electrolysis of water'

'Numbering and measuring' [in this volume]

1888, 'On the intrinsic light of the retina'

1889, 'On atmospheric movements'

1890, 'The energy of the waves and the wind'

1891, 'An attempt to extend the application of Fechner's law in the colour system'

1892, 'The principle of least action in electrodynamics'

'Electromagnetic theory of colour dispersion'

1893, 'Conclusions from Maxwell's theory as to the movements of pure ether'

1894, 'On the origin of the correct interpretation of our sensual impressions'

* * *

The second half of the 19th century was an intellectual battlefield. Earlier Hegelian and other idealist conceptions developed further, based on a conviction that attempts at critical understanding of human nature, the 'human essence', by analytic means had failed. As against such Hegelians – Helmholtz remarked on 'the exaggerations of Hegel and Schelling' – a new empiricist trend emerged among scientists along with pioneering successes in physics, chemistry, and physiology. For these philosophical scientists, the fact that human knowledge is severely limited by constraints on the senses is purely contingent, a physiological fact, to be understood, mediated, and overcome. (There were other philosophical tendencies too: a physical materialism based upon reductive explanation – Helmholtz wrote his father that he was 'not in the very least... believing in the trivial tirades of [the materialists]' – and the historical materialism which fused a naturalist interpretation of Hegel and his dialectical logic with empirical natural science and empirical studies of the processes of social history.[2] The two major trends, idealist and empiricist, traced their origins to the work of Kant, and both believed one of their major tasks to be precise clarification of their actual and, to be sure, critical relation to Kant.

Kant had considered the limiting determinants of human knowledge to be epistemological, rooted in the *a priori* character of the interpretations of human sensations. Hegel and the romantics rejected this, and the Kantian limitation was transformed, and elevated to become the center of another epistemology: for them, knowledge and its limits are not to be found in the senses. In contrast, the positivistic empiricists seemed, perhaps only superficially, to be returning to Kant. Helmholtz is often seen to be in the midst of this new empiricist-positivist current of thought. Indeed, even Maurice Mandelbaum, in his majestic *History, Man and Reason* (Baltimore, 1971), accepts the dichotomy of the romantics and the positivists; he takes as truth the occasional statements of Helmholtz, duBois-Reymond and their friends that they were genuine, 'hard-core' positivist-empiricists. For our part, we see Helmholtz' statements as complying with the image of science held by the surrounding general intellectual community, and not with his own.[3] Indeed, a number of mid-20th century presuppositions about the metaphysical, physiological, and social characteristics of human knowing condense into just such a conflation of positivism with empiricism. But to our understanding, Helmholtz was an empiricist, not a positivist, unlike Kirchhoff who was both. Instead of attempting to define positivism here, let us simply accept Gustav Kirchhoff as the archetypical scientific positivist, 19th century style.[4]

What typified Helmholtz, and duBois-Reymond and many others, was a conflict between genuine Kantian *a priori* and a deeply rooted successful scientific empiricism, the latter embracing observation, experiment, measurement, self-conscious inductive procedures, and cautious demands for empirical significance of all admissible theoretical terms and scientific laws. No doubt, their philosophical outlook and *Problemstellung* were formed by Kant; but their successes in physical, and even more in physiological and psychophysical investigations taught them that much of what Kant delegated to the *a priori* can convincingly and simply be traced to human experience in a nearly Newtonian-Euclidean world. The quarrel of empiricism with Kant was perennial for them, and apparently it was fruitful in their creative work for decades.

Helmholtz' life-long struggle is dramatically illustrated by the well-known story that the 25-year old Helmholtz wrote a strongly Kantian introduction to his epoch-making paper of 1847, 'On the conservation of force'[5] but discarded it on the advice of duBois-Reymond. His close friend warned him that the prevailing anti-metaphysical, even anti-theoretical, image of science held by scientists would reject not only such a philosophical introduction but the scientific paper as well. It might be interesting to speculate what was written in that first introduction, especially since the actual printed introduction was in fact philosophical, and later seen to be the leading methodological "foreword of modern science in the second half of the 19th century."[6] Our hunch is that most of the discarded material was included, 32 years later, in Helmholtz' centennial lecture for the University of Berlin, 'The Facts in Perception' (the last essay in this volume).

Concerning Helmholtz' relation to Kant, much was written in the late 19th and early 20th centuries, especially by neo-Kantian philosophers, but the often redundant literature has been almost forgotten. We would only suggest that 'Kant and Helmholtz' is still a point of importance, not only for the 'pure' history of ideas (i.e. the history of disembodied ideas) but even more for that new and complex discipline, the historical sociology of scientific knowledge.[7]

For the moment, consider the book before us. Hertz and Schlick selected and annotated the four papers almost as though, unaware, they were mirroring the two philosophical trends in 19th-century philosophy of science: a strong, but tacit, neo-Kantian perspective of the natural-philosophy-minded Paul Hertz in Rostock; and the incipient, but far from tacit, Vienna Circle logical positivism of Moritz Schlick, then at Göttingen. We find it of some importance for this book that the two commentators divided the essays between themselves, neither attempting consensus nor de-emphasizing their differences. They do, as they draw vital issues from these four papers, share a common problematic: their questions are those of Helmholtz, but they try to give modern answers.

As it affects us today, the Schlick/Hertz commentary suggests four stages of reflection (or, if we may stick out our common neck, five).

(i) The Kantian *Problemstellung* with Kant's answers.

(ii) The Kantian *Problemstellung* reformulated, in view of 19th-century scientific developments, with Helmholtz' answers; this, in the Schlick/Hertz commentary, is the classical Helmholtzian epistemology, of decisive influence upon the growth of science (especially but not only in Germany), and also of formative influence upon the sociological dimension of science, on the institutionalization of physics, biology, psychology and medicine, the problems tackled and methods adopted, the relation of science to applied science and technology, and – *pace* Helmholtz' multidisciplinary genius – the division of labor among the special branches of science. (We may conjecture: in the continuing critical dialogue in psychology between the Helmholtzian and say the Wundtian, if Wundt had won the upper hand, or had at least an equal influence for his alternative image of what is empirically respectable science, it is possible that the behavioral sciences in the 20th century would have been freed not only from the confining straitjacket, but also from the guiding and heuristic model, of the methodology of the natural sciences.)

(iii) The ambivalent twenties. These are typified by Hertz and Schlick in 1921, who held to Helmholtz' epistemological problems but sought new answers by repeating his own conceptual strategy: rely on Helmholtz' formulation of problems, but use 20th-century scientific knowledge in order to correct or up-date his conclusions. (Here we may observe, with respect and with caution, that they failed to *reformulate the scientific problems*; and this was the very step that made Helmholtz' contributions of fundamental, that is, epistemological importance. But Helmholtz saw himself as, and was, a working scientist *while* deeply engaged in his explicitly philosophical work. This was hardly true of Schlick and Hertz.[8] This division of labor between scientific and epistemological work is not their limitation alone, nor unknown to some of the most powerful minds among the neo-Kantians, even to Cassirer.)

(iv) The dominant metaphysics today. The anti-metaphysical scientific community in all its branches in our own time (we do not speak here of philosophers and historians of science) disregards critical epistemology. Modern physics, especially in its operational practice, molecular biology in its central dogma, behavioristic psychology in theory and in therapy, experimental-cognitive psychology, all tend to ignore philosophical criticism. (While Hertz and Schlick presaged this situation, they could not transcend it in 1921; the great intellectual experiments of 'rational reconstruction' and logical appraisal of the sciences by the Vienna positivists and the Berlin empiricists, by Carnap and Reichenbach and the rest, lay ahead. In that process, reconstruction of scientific theory always lags behind the most advanced innovative constructions of the scientists.)

(v) Novel epistemology. As an active part of research work, epistemological attention curiously creeps back slowly in unexpected areas: symbol-centered anthropology, cognitive and socializing educational studies, psychology and sociology of myth, cross-cultural investigations into ways of thinking, perceiving, numeracy, literacy, indeed general cross-cultural and historical studies of rationality.[9] (But will this re-invention of epistemology, once again in the midst of working science – although at first glance far from either physiology or physics – enlighten science in general, and even physics, in the present impasse? We do not know. At any rate, we believe it is one way to get the present value-free vs. value-based, objectivist vs. relativist, debate in the social sciences onto a constructive track. And the new 'comparative epistemology' is concerned with differing ways of dealing with nature, with cognitive *praxes*. It might be Helmholtzian.)

Accompanying the past century of development of the philosophy of science, and closely connected, are changing attitudes toward history; whether to ascribe to historical knowledge a special epistemological status or to consider it as part of the general epistemological problematic; whether to ignore historical consciousness or to place it in the center of inquiry; these views change and interact with changing images of scientific knowledge. Our stages (i) and (ii) were times when historical understanding was thought to be a special mode of thought, times when the empiricists fought a heavy battle to eliminate metaphysical 'history' from the legitimate scope of the 'sciences'. For most (Marx and Engels aside), the struggle against history coincided with the more general war against metaphysics, Hegel, and *Naturphilosophie*.

With stage (iii), modern empiricists and positivists (Zilsel and Neurath aside), systematically ignored history until very late in their game. They were seeking time-independent truths and methods, content-full and context-free; or, to follow Reichenbach's incisive and revealing formulation, they were seeking contexts of justification (rational proof) which are to be radically differentiated from contexts of discovery (psychological, social, or historical). The neo-Kantians, even Cassirer, mostly bracketed or postponed the problems of historical cognition and of the historical resonance of ideas altogether. And so, stage (iv), in the last fifty years historical studies have developed nearly independently of the natural sciences. After some intense encounters, historians even gave up any claim to the social and behavioral sciences which, for their part, seceded from the humanities (actually from history) in their understandable but empty urge to copy or adapt to the natural sciences. The conflicting claims over scientific and alternative 'humanistic' modes of understanding may be illuminated by the comparative sociology and psychology of our stage (v), perhaps mediated by the historical sociology of the natural and social sciences. We see no other epistemological synthesis between history and the sciences (natural and 'behavioral') within the horizon.[10] And we think the link to Helmholtz should be twofold: first, a methodological analogy, i.e. epistemological inquiries from within new and actual (social) science, and second, Émile Meyerson's Helmholtz-interpretation (in his notion of the *plausible*) as an early warning indication of the human character of the new synthesis.[11]

* * *

As might be expected of the author of *Identité et Réalité*, Meyerson gave closest attention to elements of persistence through change as they appear in Helmholtz' works. Meyerson's preface to the first edition (1907) of his book begins by quoting from Helmholtz in the *Physiological Optics*: "The more I study phenomena the more I am struck by the uniformity and agreement in the action of the mental processes." Meyerson immediately adds that Helmholtz means the identity of unconscious psychological processes (here those that accompany visual perception) with those of conscious thought. If with Helmholtz we understand that visual perception contains "condensed reasoning very difficult to follow", and that the physics, physiology, and psychophysics of the interdisciplinary scientist can unravel the hidden processes, then the philosopher likewise may try to bring to light *a posteriori*, by historical investigation of scientific reasoning, case by case as it were, the actual but covert and 'condensed reasonings' of the human mind in its clearest practice – modern science. Meyerson was led by this procedure to 'laws which rule the human mind', and his empirical *a posteriori* method led, as he thought, to an *a priori* heuristic principle of explanation common to all theoretical inquiry. This is a provocative neo-Kantianism beyond Helmholtz (and regretfully beyond this Introduction).

Meyerson, as he begins his text, quotes Helmholtz again, this time from the essay on the conservation of energy: "The principle of causality is nothing else than the supposition that all phenomena of nature are subject to law." By 1878, Helmholtz speaks of 'orderliness according to law'. Then Meyerson begins his critical examination. We highlight parts of his argument:

(i) causality is replaced by lawfulness (Helmholtz' term), and these are not the same;

(ii) lawfulness is not necessarily empirical generalization or mere comprehensiveness;

(iii) nor is it either logically or practically a presupposition, but rather it is a genuine conviction, indissolubly linked with primitive as well as civilized intentional human actions;

(iv) how then was conservation of energy *proved* by Helmholtz, given that he invoked "the final aim of science" as explanation through central forces with intensity varying with distance, and given his further claim (necessary to his proof) that empirical evidence proved the impossibility of perpetual motion? Surely these were *both* wrong?

(v) rational proof may be a mistaken demand; perhaps the conservation law is plainly empirical, since Helmholtz followed his proof by adding that "complete confirmation of the law must be considered as one of the principal tasks which physics will have to accomplish in the years to come"? But how impossible! – we do not know all possible forms of energy, nor do we know any relation among the known ones to show that they exhaust the possibilities, and in fact our knowledge of forms of energy has

time and again come by accident; no, our certainty exceeds the
data, our conviction exceeds the proofs;

(vi) perhaps, then, 'lawfulness' expresses properties of bodies? but
the concept of properties retains the possibility of change in
time, while laws do not.

And so Meyerson continues to debate, concluding at last, with Helmholtz, that conformity to law dominates all the sciences, and that science
is not positivist (for it contains no data to satisfy the Comtean requirement of all positivism, 'data stripped of all ontology').

* * *

As Meyerson, so also Ernst Cassirer followed Helmholtz as a guide to
mature scientific practice. In Cassirer's view, as late as 1940, in the 4th
volume of *The Problem of Knowledge*, Helmholtz led a post-Hegelian
generation with the heroic cry, 'Back to Kant'. He did this, we should
note for Hertz and Schlick, by a new link between natural science and
philosophy which depends upon physiology and psychology rather than
physics and geometry. To Helmholtz, it was Johannes Müller's discovery
of the specific energy of the sense organs that was "the empirical
realization of Kant's theoretical explanation of the nature of human
understanding". Nevertheless, how new or specifically Kantian was this
idea? The pre-critical empiricist Kant, after all, had *begun* with a physiological *a priori*. Indeed, Cassirer saw that Helmholtz was persistently
wishing to carry out an empiricist program, adapting as he did the *a
priori* nature of space 'in the Kantian sense' – he spoke of space in its
most universal form as 'the possibility of coexistence' – both to everyday
experience of psychophysical distance measurements and to the advanced
hypothetical alternatives of the non-Euclidean geometries. Similarly,
concerning arithmetic, Helmholtz linked experience with pure mathematics via the psychophysics of numbering and counting; once more he
was Kantian, on this occasion by adapting the *a priori* intuition of time to
the possibility of the act of counting. Nevertheless, even Helmholtz'
empiricism loses some of its objective reference, for his physiological
theory of signs (*Zeichen*) portrays human knowledge as a functional
correspondence with, rather than an actual similarity to, the object
known. Correspondences are between relations among signs and relations

among objects. It is relations that correspond. Cassirer expressed it in 1910: "What is retained (in the sign) is not the special character of the thing signified, but the objective relations, in which it stands to others like it.[12]

Cassirer was keenly aware of the epistemological unity of Helmholtz' varied works, and this was emphasized in Cassirer's resonant phrasing of Helmholtz' views: "What matter actually and essentially 'is' can only be expressed in the sign language of our senses".[13] Helmholtz himself, at the end of a work on psychophysics, the *Physiological Optics*, wrote: "The main thesis of the empirical standpoint is that sense impressions are signs for our consciousness, the interpretation of which is left to our intellect." His dualism of knowledge and reality is optimistic since this mediated relation between them is firm, and reality includes both the external object and the sense organs with their sense-impression states. The correspondence between relations among objective phenomena and theoretical (or practical) cognitive reasoning is genuine, in Helmholtz' view. Epistemological signs or signals, which are sense-impressions understood by ordinary intelligence or by scientific theory, are central, since "the sign... becomes [for Helmholtz] the basic concept of scientific epistemology."

Signs of interrelationships among objects, no doubt, but of what else? To Cassirer, a Kantian spirit dominates classical physics, including the interpretation of epistemological signs, and through physics all of science, in those "highest and most mature insights" which we owe to Helmholtz. The search for ever more general laws must be understood as a regulative principle, says Cassirer: it must be given *a priori* since there is no possibility of proof, and the search constitutes the causal nature of scientific explanation and of reality. Without that causal nature, there is no intellectual action, no human *praxis*. But we, and Cassirer too, know there is no Laplacean completeness or predictability in science either. Helmholtz had reassuring advice for the fearful inductivist: "Here only one counsel is valid: Trust the inadequate and act on it; then it will become a fact."[14]

* * *

At various times and circumstances, Helmholtz influenced his own and subsequent generations of scientists. E. G. Boring's judgment sets the stage (from the Preface of 1942 to his *Sensation and Perception in the History of Experimental Psychology*):

No reader of this book will need to ask why I have dedicated it to Helmholtz. There is no one else to whom one can owe so completely the capacity to write a book about sensation and perception. If it be objected that books should not be dedicated to the dead, the answer is that Helmholtz is not dead. The organism can predecease its intellect, and conversely. My dedication asserts Helmholtz' immortality – the kind of immortality that remains the unachievable aspiration of so many of us.

Among physicists, the roster of influence must include Maxwell and Kelvin, Boltzmann and Planck, and Heinrich Hertz. Of Hertz, Helmholtz wrote that he was "the one amongst all my pupils who had penetrated furthest into my own circle of scientific thought... endowed with the rarest gifts of intellect and character."[15] Among psychologists special mention must be made of Helmholtz' role in the work of Sechenov and Pavlov, the great Russian physiological psychologists. Ivan Sechenov studied with Helmholtz in his laboratory for several years about 1860, as well as with duBois-Reymond for some time; Pavlov in turn was indebted to Sechenov. Aside from the broadly physico-chemical approach to investigating psychological and neurological phenomena, Sechenov was struck by Helmholtz' idea of 'unconscious inferences' or 'unconscious conclusions' as the conjectured essential process of human cognitive and also animal perception. In 1928, in a celebrated passage in his lecture on 'Natural Science and the Brain', Pavlov wrote[16]:

Evidently what the genius Helmholtz referred to as 'unconscious conclusion' corresponds to the mechanism of the conditioned reflex. When, for example, the physiologist says that for the formation of the conception of the actual size of an object there is necessary a certain length of the image on the retina and a certain action of the internal and external muscles of the eye, he is stating the mechanism of the conditioned reflex. When a certain combination of stimuli, arising from the retina and ocular muscles, coincides several times with the tactile stimulus of a body of certain size, this combination comes to play the role of a *signal*, and becomes the conditioned stimulation for the real size of the object. From this hardly contestable point of view, the fundamental facts of the psychological part of physiological optics is physiologically nothing else than a series of conditioned reflexes, i.e. a series of elementary facts concerning the complicated activity of the eye analyzer.

Toward the end of his life and work, Helmholtz went far beyond the first law of thermodynamics, that unifying principle of the conservation of energy, interdisciplinary and universal as that was. Moreover, he went beyond the empiricist epistemology of the foundations of geometry and arithmetic. Meditating on his work on free energy, and on the second law of thermodynamics, and considering the general questions that arise concerning the relations of thermodynamics – the general science of energy phenomena – to dynamics, Helmholtz went beyond the earlier, though recent, achievements of Boltzmann, Clausius, and Rankine. Whereas they demonstrated quite generally a correspondence in formal behavior between thermodynamic entropy and the action function of a periodic mechanical system, Helmholtz related his so-called 'free energy' to the negative Lagrangian of monocyclic systems. The second law had already linked the phenomenological sciences of heat, portions of electricity, and chemical dynamics. With the successful identification of formal properties in dynamics and in applied thermodynamics, it seemed hopeful that an approach might be made to a fully universal principle governing all natural processes. Formal unification had also been achieved with Helmholtz' work on the correspondence of theoretical hydrodynamics with magnetic theory, and of acoustics with optics. In 1886, in the magisterial paper on the physical meaning of the principle of least action, continuing the series on monocyclic systems, he turned from a physics of masses in motion to a general physical science which was to be characterized by two principles: conservation of energy, and minimal action. Even more striking, only one principle is needed, since the energy principle is derivable from a general principle of least action (but, as Helmholtz showed, not *vice versa*). No wonder that Cassirer sees Helmholtz' work here as the high achievement of classical physical science at the great open road toward the modern physics of quantum mechanics and general relativity.

Helmholtz' claims were strong and simple: "All that occurs is described by the ebb and flow of the eternally undiminishable and unaugmentable energy supply of the world, and the laws of this ebb and flow are completely comprehended in the principle of least action."[17] Principles, we may add, are not laws. The principle of least action, in Helmholtz' terms (which are plain enough), is a "heuristic principle" and a "guideline in the effort to formulate laws for new classes of phenomena." To the Kantian, Cassirer, it is a 'regulative principle'. But as heuristic, such principles either show their power or fade away; as Hertz so beautifully praised Helmholtz, recognition of the scientific significance of the action principle was a genuine 'discovery'. In a perceptive account of the 1886 paper, Stephen Winters has very recently shown that Helmholtz' place in the triumphant development of Lagrangian and Hamiltonian mechanics was to answer P.G. Tait's repeated plea (and prophecy) for "... the coming of the philosopher who is to tell us the true *dynamical* bearings of varying action and of the characteristic functions." Once more, this time at the end of his life, Helmholtz was concerned with 'force', as he had been in that youthful work on conservation. Now we can see, with Winters, that the early usage of *Kraft* and its subsequent English replacement by 'energy' rather than 'force' needs to be reconsidered. After all, it was the Lagrangian 'generalized forces', subject to a principle of minimal action, that remained in classical physics; and it was a universal "interdependence and interconnection" of natural forces that would best describe the phenomena of nature. Winters reminds us that Faraday rejected Maxwell's suggestion that it would be better to speak of 'energy' than of 'force': "Faraday stubbornly refused to say anything other than what he meant." The 1847 conservation law did, we must agree, concern transformations of energy, but its discoverer's intuition was for a law of forces.

In the end a new mathematics was required and developed to express the physical insight; then the demonstration of sets of formal identities and conceptual similarities among empirically diverse physical domains brought Helmholtz to inquire into deeper physical reality. What physical quantity, if not energy, was constrained and minimized by a law of least action, or by any equivalent minimal principle? We cannot hope to decide whether science has succeeded in this program in the new century, for we wonder what Helmholtz would have said to the puzzling *new* succession of interpretations of that 'physical' quantity whose minimax conditioning was so exquisitely prescribed by him. He knew the old succession of interpretations of the minimized quantity better than anyone: Fermat, Leibniz, Maupertuis, Euler, Lagrange, Gauss, Hamilton, each seeing a different conserved and constrained property or entity, and Cassirer even reminds us of Hero of Alexandria's minimal light path. Helmholtz chose his 'kinetic potential' to be minimized for equal time intervals. But it was soon replaced, and so was so much that was classical. Helmholtz' heuristic remains.

* * *

Marxists since the time of Friedrich Engels have read Helmholtz with great interest, and for reasons similar to those of Cassirer, Meyerson, Maxwell, Pavlov, or Schlick and Hertz. Helmholtz, in his own self-critical dialectic of subjective and objective moments in scientific practice, was seeking to renew and fulfil philosophy through science. In the mid-seventies, he wrote to a friend: "I believe that philosophy will only be reinstated when it turns with zeal and energy to the investigation of epistemological processes and of scientific methods... a real and legitimate task. The construction of metaphysical hypotheses is vanity." He was against the static and abstract, against the unchanging *a priori* of uncritical and irrefutable metaphysics, against dogmatic and obscurantist thought, against vitalism and all spiritualist explanations, against anti-Semitism and against all nationalist and chauvinist prejudice in science as in ordinary life; no wonder that his thought and work would attract attention from materialists, Marxist or otherwise, and from every tendency of anti-metaphysics. A Marxist or a positivist could agree with, could write, Helmholtz' angry defense of philosophy against metaphysics: "Metaphysics... that so-called science which strives by pure thought to formulate conclusions as to the ultimate principles of the coherency of the Universe... I must protest against my objections against metaphysics being transferred to philosophy in general (since) in my opinion nothing has been so pernicious to philosophy as its repeated confusion with metaphysics." [18]

Other views, more positive, join Marxian materialists to Helmholtz. Even in his conception of sensations as signs, distant as it seems to be from any passive materialist doctrine of knowledge as reflections or copies of reality, Helmholtz should be congenial. After all, he describes knowing as the active practice of brain and sense organs; the latter provide the signs which arise by material processes of interaction with external objects; the former interprets these signs (unconsciously or consciously) in the light of experience and practice, and then tests the interpretations through deliberate efforts to produce changes in the external world by bodily and instrumental movements. These movements are not merely activities of scientific inquiry. The physiology of cognition permeates all life, and cognition is subordinate to it. Even science is subordinate to general human practice, for while Helmholtz often said that the sciences "evoke and educate the finer energies of man", he went on to say that knowledge is not the sole aim of mankind upon this earth, and that "it is in action alone that (man) finds a worthy destiny."

Nor was Helmholtz indifferent to 'the social function of science'.

He was distressed by the misunderstanding of mathematics and the natural sciences which was so widespread (even then!) among those who seemed to live in a different world, the cultured humanists with their philosophical and historical interests. Science had its methods and aims, and among their achievements were "solutions which the work of scientists offers for the great mysteries of human existence." In his extraordinary inaugural address of 1877 as Rector of the University of Berlin, he spoke openly and firmly in this vein: science for peace, science for humanity; science was not for his country alone. And these democratic slogans and outlook, these liberal views, were joined to his grateful pride in that Germany which had led the European struggle against authority, which (in Koenigsberger's summary of Helmholtz' view) "testified even unto blood for the right of such convictions, in the 16th century," and which in the 19th century was a nation with "greater fearlessness of the whole entire truth" than even England and France where scientists were subject to social and churchly prejudices.

These progressive feelings reached to democratic process too, for Helmholtz spoke on behalf of the national need to have the interests of the majority prevail, and to have a legitimate voice in the government for the "political interests of the working class." Perhaps we may describe the ennobled von Helmholtz as one who transcended his Prussian time and place to become a classic scientific and democratic humanist of modern Europe.

Engels discussed Helmholtz with care, particularly in his informal notes on current scientific progress which were published a generation later as *Dialectics of Nature*.[19] Lenin too treated Helmholtz, whom he described as "a scientist of the first magnitude", with great interest. In the end, Lenin was dissatisfied with Helmholtz' theory of cognitive signs.[20] Lenin found Helmholtz to be inconsistent, at one place a materialist about human knowledge, at another place agnostic and sceptic, and at yet other place a Kantian idealist, in sum a 'shame-faced materialist'. But these are not final judgments, nor conclusive evaluations of Helmholtz overall. Renewed interest in Helmholtz has come from Soviet and German Marxists, most rigorously and sympathetically from Klix, Kosing, Hörz and Wollgast a few years ago,[21] (and understandably motivated in part by the unending Marxist coming-to-grips with the logical empiricist philosophy of science, which is to say, in

Helmholtz' case, with Schlick and Hertz). Nevertheless, a full Marxist appreciation of Helmholtz has not been written. After all, Koenigsberger was right to bring Helmholtz' epistemological situation to a blunt conclusion:[22]

> "... all is not light that is perceived as light...
> ... there is also light to which we are not sensible...
> ... The radiation which we term now light, now radiant heat, impinges on two different kind of nerve end-organs, in the eye and in the skin, and the disparity in quality of the sensation is due not to the nature of the object sensed, but to the kind of nervous apparatus that is thrown into activity..."

> "From this simple and obvious truth, Helmholtz developed his entire theory of knowledge."

To us, it seems time that the Marxists, and not only the Marxists, must embed this scientifically "simple and obvious" theory of human knowing within a broad and equally scientific theory of the historical and cultural practices of humankind, "thrown into activity...".

* * *

This edition of Helmholtz' essays was prepared by the devoted and intelligent labor of Malcolm F. Lowe as translator and interpreter of 19th-century German conceptual vocabulary into 20th-century English. Other translations of these writings are noted in the Bibliography, but none seem to us to bear the burdens of faithful accuracy and dependable lucidity as well as Lowe's work. His is the first translation of the Schlick/Hertz annotations.

We have added a Bibliography. There is (A) a list and guide to Helmholtz' works, including English translations known to us; (B) a short list of published biographical materials; (C) a substantial list of works on Helmholtz, with emphasis upon those of philosophical or historical relevance, including biographies; (D) a bibliographical index to works cited either by Helmholtz, or by Hertz and Schlick in this book, together with English translations where known. We have taken particular care in Part C to include studies published since 1921 in English, German, French or Russian, but we would be grateful to be informed of publications that we have missed. One forthcoming noteworthy publication of interest to readers of this book is Henk L. Mulder and Barbara Van de Velde-Schlick's edition in English of Schlick's *Philosophical Papers 1910–1936* (Vienna Circle Collection: Reidel, Dordrecht and Boston, in preparation).

ROBERT S. COHEN　　　YEHUDA ELKANA
Boston University　　　*The Hebrew University of Jerusalem*

NOTES

[1] See Helmholtz, *Wiss. Abhand.*, Bibliography Part A.

[2] On physical materialism, see Frederick Gregory, *Scientific Materialism in 19th Century Germany* (Dordrecht and Boston, 1977).
On the relation of historical materialism to science, see R.S. Cohen, *Karl Marx* and *Friedrich Engels* in *Dict. Sci. Biog.*, supp. vol. (New York, 1977).

[3] There is a rich literature on nineteenth century philosophical trends, especially epistemology and theories of perception, both as a branch of physiology and as a branch of psychology. On images of knowledge among laymen, general intellectuals and specialists, see Y. Elkana, 'The Problem of Knowledge in Historical Perspective', *Proc. Second International Humanistic Symposium* (Delphi, 1972) pp. 191–247. See also a review-discussion of Mandelbaum's book by Leon Pompa, *Inquiry* 16 (1973) pp. 323–351.

[4] On modern positivism in the philosophy of science, several accounts may be mentioned (from a vast literature) as helpful:
(1) *Wissenschaftliche Weltauffassung: Der Wiener Kreis* [translated as 'The Scientific Conception of the World: The Vienna Circle' in Otto Neurath, *Empiricism and Sociology* (Dordrecht and Boston, 1973)].
(2) Richard von Mises, 'Ernst Mach and the Empiricist Conception of Science' *Einheitswissenschaft* 7 ('s-Gravenhagen, 1938) as translated in *Ernst Mach: Physicist and Philosopher*, ed. R.S. Cohen and R.J. Seeger (*Boston Studies in the Philosophy of Science*, vi, Dordrecht and New York, 1970) pp. 245–270.
(3) W. Stegmüller, *Main Currents in Contemporary German, British and American Philosophy* (Dordrecht, 1969), ch. 7 'Modern Empiricism: Rudolf Carnap and the Vienna Circle'.
(4) On the distinction between positivist and non-positivist empiricism, (and the relation of both to materialism), see R.S. Cohen, 'Dialectical materialism and Carnap's logical empiricism', *The Philosophy of Rudolf Carnap*, ed. P.A. Schilpp (Lasalle, Illinois, 1963) pp. 99–158.

[5] The memoir was first translated into English in 1853. For our own discussion, see Yehuda Elkana, 'Helmholtz' *Kraft*: An Illustration of Concepts in Flux' [as in Bibliography C]; other literature is discussed below, and cited in the Bibliography.

[6] The phrase is Koenigsberger's in his *Hermann von Helmholtz*, Welby tr., p. 38 [as in Bibliography B].

[7] Major sources on this issue are found in Bibliography C, e.g. Cassirer, Erdmann, Goldschmidt. A recent study by one of us [Elkana, *The Discovery of the Conservation of Energy* (London and Cambridge, 1974)] deals mainly with the essential tension

between empiricism rooted in physics, physiology and psychology and the metaphysical heritage of Kant and even of Fichte.

[8] Hans Reichenbach wrote perceptively on this point in his contribution to *Albert Einstein: Philosopher-Scientist*, ed. P. A. Schilpp (Evanston, 1949) pp. 290–291. His words might apply to Helmholtz as well as to Einstein, it seems to us:
"The analysis of knowledge has always been the basic issue of philosophy; and if knowledge in so fundamental a domain as that of space and time is subject to revision, the implications of such criticism will involve the whole of philosophy... Einstein's primary objectives were all in the realm of physics. But he saw that certain physical problems could not be solved unless the solutions were preceded by a logical analysis of the fundamentals of space and time, and he saw that this analysis, in turn, presupposed a philosophic readjustment of certain familiar conceptions of knowledge... It is this positivist, or let me rather say, empiricist commitment which determines the philosophical position of Einstein..."

[9] Some promising beginnings may be mentioned:
M. Cole and S. Scribner, *Culture and Thought* (New York 1974).
R. Horton and R. Finnegan, ed., *Modes of Thought* (London, 1973).
B. Wilson, ed., *Rationality* (Oxford, 1970).
McCole, Gray, Glick and Sharp, *The Cultural Context of Learning and Thinking* (London, 1971).
Robin Horton's series of papers on African science in *Africa*.

[10] On the 19th-century attitude toward historical knowledge, see Hayden White, *Metahistory: The Historical Imagination in Nineteenth Century Europe* (Baltimore, 1973).

[11] On this, see Meyerson in Bibliography C.

[12] Cassirer, *Substance and Function*, p. 304.

[13] Cassirer, *Determinism and Indeterminism in Modern Physics*, p. 130.

[14] As in Cassirer, *ibid.*, p. 63; see also 'The Facts in Perception' in this volume.

[15] See Helmholtz' 'Preface' (unpaginated) to Heinrich Hertz, *The Principles of Mechanics*, tr. Jones and Walley, new introduction by R. S. Cohen (Dover reprint, N.Y., 1956).

[16] Ivan P. Pavlov, *Lectures on Conditioned Reflexes* (New York, 1928) pp. 126–127.

[17] From the posthumous text on the history of the principle of least action (see Bibliography A), as translated in Cassirer, *ibid.*, p. 49.

[18] Koenigsberger, *ibid.*, p. 427; phrases quoted in the following several paragraphs derive from Koenigsberger, pp. 307–308.

[19] On Engels and Helmholtz, see entries for Engels, Klare, Reiprich in Bibliography C, and Cohen, *Friedrich Engels, op. cit.*

[20] See Lenin, *Materialism and Empirio-Criticism* (Bibliography C), *passim* but especially ch. 4, section 6.

[21] Particularly in the sessions of the DDR Academy of Sciences devoted to Helmholtz in 1972 (see Klare *et al.*, Bibliography C), and in Wollgast's paper at the Rostock symposium of 1970 on Moritz Schlick. Soviet interest has been expressed primarily by historians of science. There appear to be no writings on Helmholtz by 'Western' Marxists, except for brief statements, e.g. J. D. Bernal, *Science and Industry in the 19th Century* (London, 1953), p. 63. On the general social and economic background there are many Marxist writings, those on conservation of energy being relevant to Helmholtz, e.g., S. Lilley 'Social Aspects of the History of Science', *Arch. Internationales d'Hist. des Sciences* **28** (1948) pp. 326–443, and L. Rosenfeld, 'La Genèse des principes de la thermodynamique' *Bull. de la Soc. Roy. des Liège* (1941) pp. 199–212, [Eng. tr. in *Selected Papers of Léon Rosenfeld*, ed. R. S. Cohen and J. J. Stachel (*Boston Studies in the Philosophy of Science*, xxi, Dordrecht and Boston, 1977)].

[22] Koenigsberger, *ibid.*, p. 98.

TRANSLATOR'S NOTE

As Helmholtz has suffered in some previous translations[†] it is hoped that this one of some of his key papers, together with the valuable commentary of Hertz and Schlick, will go some way towards redressing the balance. Anyone who cares to compare this translation (and others) with the original will be able to judge how closely I have endeavoured to follow both his words and his spirit. However, I do not flatter myself that an ideal solution has been found to every problem, and any suggested amendments will be gratefully received and examined.

At the same time, it should be noted that some of Helmholtz' writings exist in more than one version. I have not attempted a complete collation of different versions, but strictly followed the text given by Hertz and Schlick, except where errors required correction[††]. Occasionally these have been explicitly noted.

Certain difficulties encountered in the translation require special mention. In the first place, there is the question of Kantian terminology, since Kant's philosophy is a major topic for Helmholtz' attention. There exists a more or less standard system of translating this terminology, as exemplified in Kemp Smith's version of the *Critique of Pure Reason*. However, some of the English equivalents used are highly artificial, chosen for the purposes of professional philosophers, whereas Helmholtz expressly did not always write with such an audience primarily in mind.

Two groups of Kantian terms may be given as examples. Firstly, the family of words 'Anschauung', 'anschaulich', etc., has also been given here the standard translation 'intuition', 'intuitive', etc. However, I have used where possible especially 'intuitive(ly)', because even someone with little knowledge of Kant can acquire some understanding of Helmholtz' use of these terms from their ordinary English use in such

---

[†] The faults have varied from systematic disregard of subtleties of expression, thus blunting the impact of Helmholtz' lively and imposing style, to incredibly gross errors.

[††] They confess to having been pressed for time in completing the volume. Their version of the text has therefore been compared with the appropriate originals.

phrases as ' an intuitive conception ', ' grasping something intuitively '.

For Kant, of course, ' intuition ' strictly means the mind's faculty for receiving data (or means something thus received), as opposed to the 'understanding', its faculty for supplying concepts under which these data are subsumed. Kant considers that all human 'intuition' is 'sensible', meaning that its possible varieties are determined by the senses. (It is in this respect that one has contact with the nonphilosophical use of 'an intuitive conception' to mean one that is easily pictured or otherwise imagined.) However, the most fundamental features of human intuition, its being always temporal and usually also spatial, are for Kant not data of sensation, but rather a framework supplied by the mind for the reception of sensations. The basic features of this framework (which he takes to include arithmetic and three-dimensional Euclidean geometry) are thus on this view necessary truths independent of experience, because only through this framework can we *have* experience.

The second group of terms, namely ' sich vorstellen ', ' Vorstellung ' etc., presented a more complex problem. The ordinary meaning of ' sich vorstellen ' is to ' imagine ', as when we talk of "imagining what would happen if…", i.e. forming mental pictures of hypothetical events. In philosophical usage, the meaning of ' sich vorstellen ' is extended to cover the formation of mental pictures in general, whether these arise from imagination in the strict sense, or in memory, or immediately through sensation, or whatever. The derived noun ' Vorstellung ' then means ' idea ' in the sense of the philosopher Berkeley.

Although ' imagine ' and ' idea ' are related to ordinary English usage, they do not display the etymological connexion between the German words involved. Consequently, the verb ' represent ' and the noun ' representation ' have traditionally been chosen as standard – but artificial – renderings.

The problem is complicated further by the fact that Helmholtz did not write all of the papers in this volume for the same audience. The opening essay was originally composed as a popularization of the issues of non-Euclidean geometry, intended for anyone with a fair high school education, thus not people who need have a professional acquaintance with the intricacies of Kant's philosophical works. But the last paper in this collection is directed especially at professional philosophers as a concise statement of Helmholtz' position as against Kant's.

So I have adopted the somewhat unorthodox solution of using different translations in the two places. In the opening paper, ' sich vorstellen ' is translated ' imagine ', and ' Vorstellung ' (which occurs comparatively rarely) as 'imagination' or 'idea', while in the last paper, standard English Kantian terminology has been used throughout. (This includes the use of ' appearances ' for ' Erscheinungen ', although the natural English equivalent is 'phenomena'. However, 'sinnlich' is mostly translated 'sensory' rather than 'sensible', and 'Bestimmung' as 'specification' rather than 'determination'.)

This approach is rendered especially apt by the following facts. The subject of the first paper is trying to conceive what it would be like to live in non-Euclidean worlds. Since Helmholtz did not suppose ours to be such a world, this was for him a question precisely of 'imagining' in the restricted (non-philosophical) sense. On the other hand, the principal statements of this paper are repeated in the final paper, including Helmholtz' key definition of ' sich vorstellen ' for his own purposes. Moreover, I have avoided using ' imagine ', ' idea ' and ' represent ' elsewhere, so that the reader may make substitutions as he thinks fit.

Thus it is hoped that the various papers will be accessible to the same audiences of English readers as Helmholtz intended amongst his German readers.

A different problem is presented when Helmholtz himself relies upon etymological connexions between words. This occurs above all (and very extensively) in the third paper. Here, even in the title the word 'Zählen' has to be translated 'numbering' (rather than 'counting'), because he uses it in close connexion with 'Zahl' ('number'). Most striking of all is his use of the family of words derived from 'gleich' ('equal', 'alike', 'the same'). This German word occurs in many compounds where in English the corresponding words are of Latin (' comparison ', ' simultaneous ', ' equilibrium ') or Greek (' homogeneous ') origin, and (as the Latin examples show) it does not always even correspond to the same component in words originating in the same language.

The crucial connexion upon which Helmholtz relies turns out to be that between ' gleich ' and ' vergleichen ' (' compare '), and clearly the point of what he is trying to do would be veiled from the English reader if one failed to preserve this connexion. So I have translated ' gleich ' as ' alike ' (to be understood in the sense ' *exactly* alike ') or ' like ' (in ex-

pressions such as ' of like kind ') as far as possible, while ' vergleichen '
has been translated ' liken '. This is somewhat artificial, in that to ' liken '
normally means in English to compare things whose similarity is not
immediately obvious, whereas no restriction of this kind is implied by
the German word. On the other hand, the German word itself does not
normally mean to ' make alike ', although it is this etymological implication on which Helmholtz seems to play.

Consistency of translation has also been maintained throughout the
book, as far as practicable, for numerous other words, which there is no
special reason to list here.

In the case of all publications cited by a non-English title, an English
translation (usually as literal as possible) has been appended both as an
indication of the contents and to facilitate identification of any English
translation which has occurred or may occur. I note finally that *asterisks*
are used for footnotes in the original text, *daggers* for footnotes which I
have added myself in translation, and *square brackets* to enclose additions
to the text (whether made by Hertz and Schlick or in translation).

M. F. LOWE

FOREWORD

[1921]

It is an often stated truth that a great man's own works are his final memorial. We cannot better and more worthily celebrate the memory of a genius than to enjoy his creations and promote their influence in our own age. So it seemed appropriate to mark Helmholtz' hundredth birthday by keeping alive the memory of this "intellectual giant", as Maxwell called him, with the production of a new edition of his writings of continuing influence. [The Schlick–Hertz edition appeared in 1921.]

His complete works could not be considered for this purpose. The two great classics on physiological optics and acoustics will have to continue to stand apart individually as lofty monuments to his creativity. On the other hand, his numerous lesser writings mostly deal with issues of physics and physiology which no longer have any very close relationship to the modern problems in these domains, as in both sciences – especially in physics – such striking advances have been registered during recent decades, that many issues of Helmholtz' epoch have now sunk below the horizon of a changed range of vision.

Precisely in the domain of physics, the area of his greatest talents, his most ardent interests and his most brilliant achievements, it is unmistakably evident that this extraordinary man was more of a perfector than a stimulator and pioneer. His classic works on physics – one may for example recall the energy principle or his work on vortex motions – stand at the *end* of lines of development. They may thus give a secure foundation for possible fresh developments, but modern physics, though gratefully accepting the foundations created in the "classical" era, must above all search for solutions to its own new burning issues, and looks upon these writings as something lying outside the range of its own problems. It could naturally be that some future physics will have to turn again to Helmholtz' conceptions and bring seeds hidden within them to fruition. But his work seems for science at its present stage to be something completed and left behind.

There is only one field, but a very important one, to which this does not apply: it does not hold for Helmholtz' epistemological studies. The problems which Helmholtz dealt with as an epistemologist do not merely project from his time into our own, but have today in part been aroused to a new and more intensive life, because of their acquiring a significance which was then not yet foreseen. They are namely the problems of the ultimate foundations of mathematical knowledge, which Kant already felt to contain the key to the problem of knowledge in general, and which today are of much more central interest than they were in Helmholtz' time.

Helmholtz encountered these issues from the aspect which concerns the physiology of the senses, Riemann from the mathematical aspect, while modern research encountered them starting from the physical aspect. It is in this last way that their great philosophical significance has become most clearly apparent at the present time, and aroused general attention. Yet Helmholtz was already fully conscious of it, and he correspondingly made a point of disseminating the results of these "metamathematical"[†] investigations amongst wider circles through popular lectures. Even today his presentation is still eminently suitable as an introduction to these matters, and for making accessible the range of issues of non-Euclidean geometry, issues which in the modern theory of relativity have acquired such direct significance for the understanding of the physical structure of the universe. The general theory of relativity has brought a magnificent confirmation of the real significance of Riemann's and Helmholtz' trains of thought, which were formerly exposed to considerable hostility and which have often enough been portrayed, even up to the present, as an empty play with concepts, without possible application to the actual world.

Thus it is above all through Helmholtz' writings on the theory of geometrical knowledge, that the ideas of this outstanding mind can be kept alive in contemporary thought. They had to take up a large part of this volume. Helmholtz dealt with these problems in close connexion with certain other epistemological issues, which form one of the most important ways of access to philosophy, namely the issues of the theory of perception. His clear and penetrating statements on this theme are

---

† [Helmholtz' use of the term 'metamathematical' is not related to the current one.]

not unjustly counted as classic presentations, which will still continually
be referred to in the future. This aspect too of Helmholtz' creativity had
to be considered as fully as possible in our edition, if this was to fulfil its
purpose and offer a selfcontained whole.

Even if his hundredth birthday had not given an external occasion
for a summary edition of Helmholtz' epistemological studies, it would
be timely for internal reasons. If modern natural science sees itself
compelled, by the logical universality of its principles, to resume the
closest constact with philosophy, Helmholtz here anticipated it in an
unphilosophical age. His name symbolises a fruitful union of science
with an epistemologically oriented philosophy, of which he said in a
famous passage (*Vorträge und Reden*, 5th ed., vol. I, p. 88) that its only
intention was "to examine the sources of our knowledge and the degree
of its justification, a task which will always be the preserve of philosophy
and which no era can evade with impunity".

Thus, stimulated by Dr. Berliner, the editor of *Naturwissenschaften*,
to whom thanks are due, the editors gladly undertook the task whose
outcome is now presented in this volume.

The supreme criterion guiding our selection and elucidation of Helmholtz' writings was to have regard to the *topic* involved, the problems
under discussion. We are not concerned with historical or philological
questions, thus not with tracing the chronological development of
Helmholtz' views, nor with critically sifting the literature on Helmholtz,
nor either with exhaustively reproducing or enumerating all of the
passages where Helmholtz discussed some given epistemological issue.
Instead, his writings are here offered anew for the sake of the truth that
is contained in them, or can be gained from them.

Helmholtz frequently dealt with one and the same theme a number
of times, and sometimes so uniformly as to repeat, word for word,
lengthy statements from one presentation in another. For this reason
alone, the indiscriminate reprinting of all of the epistemological passages
in his works would not have served the purpose of this volume. After
examining the whole of the material, we selected the papers now put
together here, because they contain the most complete and also most
fully clarified and rounded off presentation of Helmholtz' thoughts on
epistemology. Only at relatively few points was it necessary to adduce
passages from other writings of Helmholtz, in order to give a truly

complete picture of his epistemology. Such passages have been reproduced and put to use in our comments.

Our commentary too was governed purely by the criterion of the topic involved. As complete and faithful a reproduction and elucidation of Helmholtz' opinions as possible was not striven after for its own sake, but serves throughout the purpose of knowledge and truth alone. For this reason, mere remarks had to be supplemented by amplifications and even occasional corrections.

It was above all important to relate Helmholtz' thoughts to the present state of research. Many remarks were the occasion of taking a look at modern problems, and it conformed precisely with the purpose of this book that we should not avoid such points of contact with present issues. However, we hope that the comments will not be found to take up too much room in relation to the text, but that they will rather enhance than diminish the effect of the master's words.

It has been our endeavour to make this volume basically intelligible to the *general public*, that is, it is essentially directed at a circle of readers of the same kind as the one that Helmholtz' popular addresses and lectures presupposed. But it would not have been justified to leave out everything that could only be understood by professionals, just so as not to violate the principle of popularity. The purpose of this edition required rather that one should also go into the rigourously scientific foundation given by Helmholtz. This is why, for example, the short mathematical paper 'On the Facts Underlying Geometry' was also included. Its publication in this context seemed all the more requisite, in view of the extraordinary interest encountered by H. Weyl's recent new edition of the study by Riemann on the same theme.

In our remarks too, at a number of points, it was necessary to go beyond the needs of the non-professional; thus there are certain comments (even on the popular writings) which are intended more for the expert. However, it was not necessary to distinguish in some special way the non-popular parts of the book, since the reader who is not interested in particular issues will naturally discover them immediately in any case, and can disregard them.

Each of the two editors has placed his initials under those parts of the commentary for which he assumes responsibility (M. S. for I and IV, P. H. for II and III). Yet this is not meant to indicate a strict division of work. We have rather worked together, as far as it was possible by correspondence, and exchanged views on particular issues. However, external reasons – especially the short time available, and internal ones – stemming from occasional divergences of our epistemological opinions, seemed to indicate a division of responsibility in the way stated. If for this reason one or another issue is brought up twice, this should not detract from the purpose of the book, but be rather of advantage, as the same topic always becomes clearer when illuminated from different aspects.

The numbering of the comments starts afresh with each paper. Wherever a remark on another paper is cited, this is therefore always referred to by a Roman numeral indicating the essay concerned (e.g. note II.25).

On particular issues we have occasionally been helped by friendly advice from Geheimrat F. Klein, Dr. Bernays and Dr. Behmann of Göttingen, Professor Engel of Giessen and Professor Katz of Rostock. We wish to express our sincere thanks to them here, as well as to Geheimrat C. Stumpf of Berlin, through whose good offices we were able to use, for this present edition, B. Erdmann's posthumous work *Die philosophischen Grundlagen von Helmholtz' Wahrnehmungstheorie* [The philosophical foundations of Helmholtz' theory of perception] before its publication in the *Abhandlungen der preussischen Akademie der Wissenschaften*.

We hope the guide lines followed in our edition will prove convenient in using the book, whether for private reading or for exercises in philosophical seminars. May both scholars and students always continue to be able to find a rich source of intellectual profit and pleasure in their dealings with Helmholtz the epistemologist!

P. H. & M. S.

Göttingen and Rostock, 1921

CHAPTER I

ON THE ORIGIN AND SIGNIFICANCE OF THE AXIOMS OF GEOMETRY

(*Vorträge und Reden*, 5th ed., vol. II, pp. 1–31)

The fact that a science like geometry can exist, and can be built up in the way it is, has necessarily demanded the closest attention of anyone who ever felt an interest in the fundamental questions of epistemology. There is no other branch of human knowledge which resembles it in having seemingly sprung forth ready-made, like a fully armed Minerva from the head of Jupiter, none before whose devastating aegis dispute and doubt so little dared to lift their eyes. In this it wholly escapes the troublesome and tedious task of gathering empirical facts, as the natural sciences in the narrower sense are obliged to, but instead the form of its scientific procedure is exclusively deduction[1]. Conclusion is developed from conclusion, and yet nobody in his right mind ultimately doubts that these geometrical theorems must have their very practical application to the actuality surrounding us. In surveying and architecture, in mechanical engineering and mathematical physics, we all constantly calculate the most varied kinds of spatial relationships in accordance with geometrical theorems. We expect the issue of their constructions and experiments to be subject to these calculations, and no case is yet known in which we were deceived in this expectation, provided we calculated correctly and with sufficient data.

Thus in the conflict over that issue which forms, as it were, the focus of all the oppositions between philosophical systems, the fact that geometry exists and achieves such things has always been used to prove, as an impressive example, that knowledge of propositions of real content is possible without recourse to a corresponding basis taken from experience. In answering especially Kant's famous question "How are synthetic *a priori* propositions possible?", the axioms of geometry probably constitute the examples which seem to show most evidently, that synthetic propositions *a priori* are in general possible[2]. The circumstance that such propositions exist, and necessarily force our assent, is moreover for him a proof that space is a form, given *a priori*, of all outer intuition[3]. By that he seems to mean not merely that this form given *a priori* has the character of a purely formal scheme, in itself devoid of any content[4], and into which any arbitrary content of experience would fit. Rather, he seems also to include certain details in the schema whose effect is precisely that only a content restricted in a certain lawlike way can enter it, and become intuitable for us[*].

It is precisely this epistemological interest of geometry which gives me the courage to speak of geometrical matters in a gathering of whose members only the smallest proportion have penetrated more deeply into mathematical studies than their school days required. Fortunately, the knowledge of geometry which is normally taught in high school will suffice to convey to you the sense, at least, of the propositions to be discussed in what follows.

My intention is namely to give you an account of a recent series of interconnected mathematical studies, which concern the axioms of geometry, their relations to experience, and the logical possibility of replacing them with others. The relevant original studies of the mathematicians are, in the first instance, intended only to provide proofs for the experts in a field which demands a higher capacity for abstraction than any other, and are fairly inaccessible to the non-mathematician. So I shall try to give even such a person an intuitive conception of what is involved. I hardly need remark that my discussion is not meant to give any proof of the correctness of the new insights. Whoever seeks this must simply take the trouble to study the originals.

If a person once enters geometry, in other words the mathematical theory of space[5], by the gates of its first elementary propositions, he faces on his further journey the unbroken chain of conclusions of which I spoke already. Through them, successively more numerous and complicated spatial forms receive their laws. But in those first elements are

---

[*] In his book *Über die Grenzen der Philosophie* ["On the limits of philosophy"], W. Tobias claims that some previous utterances of mine to a similar effect were a misunderstanding of Kant's opinion. But Kant specifically quotes the propositions that the straight line is the shortest (*Kritik der reinen Vernunft* ["Critique of pure reason"], 2nd ed., p. 16), that space has three dimensions (p. 41) and that between two points only one straight line is possible (p. 204), as ones which "express the conditions of sensible intuition *a priori*". Whether these propositions are given originally in spatial intuition, or whether this only gives the starting points from which the faculty of understanding can develop such propositions *a priori* – this being what my critic attaches importance to – is quite irrelevant here.

laid down a few propositions concerning which even geometry declares it cannot prove them, but can only count on the recognition of their correctness by whoever understands their sense. These are the so-called axioms of geometry.

If we call the shortest line that can be drawn between two points a *straight line*, then there is firstly the axiom that there can only be one straight line between two points, and not two different ones. It is a further axiom that a *plane* can be placed through any three points of space that do not lie in a straight line, namely a surface which wholly includes every straight line connecting any two of its points. Another and a much-discussed axiom asserts that through a point lying outside a straight line only one single straight line can be drawn parallel to the first, and not two different ones[†]. *Parallel* is what one calls two lines lying in the same plane which never meet, however far they are extended. Besides this, the axioms of geometry also assert propositions which specify the number of dimensions of both space and its surfaces, lines and points, and which elucidate the concept of the continuity of these formations. Such are the propositions that the limit of a body is a surface, the limit of a surface is a line, the limit of a line is a point, and a point is indivisible. And also the propositions that a line is described by the motion of a point, a line or a surface by the motion of a line, a surface or a body by the motion of a surface, but only another body by the motion of a body[6].

Where do such propositions come from, being unprovable and yet indubitably correct in the domain of a science in which everything else has submitted to the authority of inference[7]? Are they an inheritance from the divine source of our reason, as the idealist philosophers[8] believe, or have earlier generations of mathematicians merely not yet had sufficient ingenuity to find the proof? Every new disciple of geometry, coming to this science with fresh zeal, naturally tries to be the fortunate one who outdoes all his predecessors. It is also quite proper that everyone should have a try at it afresh, since in the situation to date one could only convince oneself by one's own fruitless attempts that the proof was impossible[9]. Unfortunately, there still arises the occasional obsessive searcher who entangles himself for so long and so

[†] [This is strictly speaking not Euclid's postulate (axiom) of parallels, but a common substitute for it.]

deeply in intricate reasonings that he can no longer discover his mistakes, and believes he has found the solution. The axiom of parallels especially has evoked a large number of apparent proofs.

The greatest difficulty in these investigations has been, and still is, that so long as the only method of geometry was the intuitive method taught by Euclid[10], it was all too easy to intermix results of everyday experience, as apparent necessities of thought, with the logical development of concepts. Proceeding in this way, it is in particular extraordinarily difficult to ascertain that one is nowhere involuntarily and unknowingly helped, in the steps which one successively prescribes for the demonstration, by certain very general results of experience, which have already taught us practically that certain prescribed parts of the procedure can be carried out.

In drawing any subsidiary line in any proof, the well-trained geometer always asks whether it will indeed always be possible to draw a line of the required kind. Constructional tasks, as is well known, play an essential role in the system of geometry. Viewed superficially, these seem to be practical applications which have been inserted as an exercise for pupils. But in truth they ascertain the existence of certain structures. They show that points, straight lines or circles such as those whose construction is required as a task, are either possible under all conditions, or characterise the exceptional cases that may be present.

It is essentially a point of this kind, about which the investigations to be discussed in what follows turn. In the Euclidean method, the basis of all proofs is demonstration of the congruence of the relevant lines, angles, plane figures, bodies, and so on. In order to make the congruence intuitive, one imagines the relevant geometrical structures moved up to each other[11], naturally without changing their form and dimensions. That this is in fact possible, and can be carried out, is something we have all experienced from earliest youth onwards. But if we want to erect necessities of thought upon this assumption, that fixed[†] spatial structures can be moved freely without distortion to any location in space, then we must raise the question of whether this assumption involves any presupposition which has not been logically proved. We shall soon see below that it does in fact involve one, and indeed one of far-reaching

---

† [Helmholtz' use of 'fest' ('fixed') instead of 'starr' ('rigid') is alluded to in note 31 below.]

implications. But if it does, then every congruence proof is supported by a fact drawn only from experience[12].

I bring up these considerations here in order, in the first place, only to make clear what difficulties we stumble upon, when we analyse fully all of the presuppositions made by us in using the intuitive method. We escape them, if in our investigation of basic principles we employ the analytic method[13] developed in modern calculative geometry. The calculation is wholly carried out as a purely logical operation[14]. It can yield no relationship between the quantities subjected to the calculation, which is not already contained in the equations forming the starting point of the calculation. For this reason, the mentioned recent investigations have been pursued almost exclusively by means of the purely abstract method of analytic geometry.

Yet it is possible besides this to give to some extent an intuitive conception of the points at issue, now that they have been made known by the abstract method. This is best done if we descend into a narrower domain than that of our own spatial world. Let us think of intelligent beings, having only two dimensions, who live and move in the surface of one of our solid bodies – in this there is no logical impossibility. We assume that although they are not capable of perceiving anything outside this surface, they are able to have perceptions, similar to our own, within the expanse of the surface in which they move. When such beings develop their geometry, they will naturally ascribe only two dimensions to their space. They will ascertain that a moving point describes a line and a moving line a plane, this being for them the most complete spatial structure of their acquaintance. But they will as little be able to have any imagination of a further spatial structure that would arise if a surface moved out of their surface-like space, as we are of a structure that would arise if a body moved out of the space known to us.

By the much misused expression 'to imagine' † or 'to be able to think of how something happens', I understand[15] that one could depict the series of sense impressions which one would have if such a thing happened in a particular case. I do not see how one could understand anything else by it without abandoning the whole sense of the expression. But suppose no sense impression whatsoever is known that would

† [See Translator's Note.]

relate to such a never observed process as for us a motion into a fourth dimension of space, or for the surface beings a motion into the third dimension known to us. Then no such 'imagining' is possible, just as little as someone absolutely blind from youth will be able to 'imagine' colours, even if he could be given a conceptual description of them.

The surface beings would besides also be able to draw shortest lines in their surface space. These would not necessarily be straight lines in our sense, but what in geometrical terminology we would call *geodetic* lines of the surface on which they live, ones which will be described by a taut thread applied to the surface and able to slide freely upon it. In what follows, I shall permit myself to term such lines the *straightest* lines[16] of the relevant surface (or of a given space), in order to emphasise the analogy between them and the straight line in a plane. I hope this intuitive expression will make the concept more accessible to my non-mathematical listeners, but without causing confusions.

If moreover beings of this kind lived in an infinite plane, they would lay down precisely our planimetric geometry. They would maintain that only *one* straight line is possible between two points, that through a third point lying outside it only one line parallel to the first can be drawn, that furthermore straight lines can be extended infinitely without their ends meeting again, and so on. Their space might be infinitely extended. But even if they encountered limits to their motion and perception, they would be able to imagine intuitively a continuation beyond those limits. In imagining this, their space would seem to them to be infinitely extended just as ours does to us, although we too cannot leave our earth with our bodies, and our sight only reaches as far as fixed stars are available.

But intelligent beings of this kind could also live in the surface of a sphere. For them, the shortest or straightest line between two points would then be an arc of the great circle through the points in question. Every great circle through two given points is divided thereby into two parts. When their two lengths are not equal, the shorter part is certainly the unique shortest line on the sphere between these two points. But the other and greater arc of the same great circle is also a geodetic or straightest line, meaning that each of its smaller parts is a shortest line between its two endpoints. Because of this circumstance, we cannot simply identify the concept of a geodetic or straightest line with that of a shortest line. If moreover the two given points are endpoints of the same diameter of the sphere, then any plane through this diameter intersects the surface of the sphere in semicircles, all of which are shortest lines between the two endpoints. So in such a case there are infinitely many shortest lines, all equal to each other, between the two given points. Accordingly, the axiom that only one shortest line exists between two points would not be valid, for the sphere dwellers, without a certain exception.

Parallel lines would be quite unknown to the inhabitants of the sphere. They would maintain that two arbitrary straightest lines, suitably extended, must eventually intersect not in just one point, but in two. The sum of the angles in a triangle would always be greater than two right angles, and would increase with the area of the triangle. For just that reason, they would also lack the concept of geometrical similarity of form between greater and smaller figures of the same kind, since a greater triangle would necessarily have different angles from a smaller one[17]. Their space would be found to be unbounded, yet finitely extended, or at least would have to be imagined to be such.

It is clear that the beings on the sphere, though having the same logical capabilities, would have to lay down a quite different system of geometrical axioms from what the beings in the plane would, and from what we ourselves do in our space of three dimensions. These examples already show us that beings, whose intellectual powers could correspond entirely to our own, would have to lay down different geometrical axioms according to the kind of space in which they lived.

But let us go further, and think of intelligent beings existing in the surface of an egg-shaped body. Between any three points of such a surface one could draw shortest lines, and so construct a triangle. But if one tried to construct congruent triangles at different locations in this surface, it would be found that the angles of two triangles having equally long sides would not turn out to be equal. A triangle drawn at the pointed end of the egg would have angles whose sum differed more from two right angles, than would a triangle with the same sides drawn at the blunt end. It emerges from this, that even such a simple spatial structure as a triangle, in such a surface, could not be moved from one location to another without distortion. It would be found equally, that if circles of equal radii (the length of the radii being always measured by shortest lines along the surface) were constructed at different locations in such a surface, their periphery at the blunt end would turn out to be greater than at the pointed end.

It also follows from this, that it is a special geometrical property of a surface to be such that the figures lying in it can be displaced freely without altering any of their angles or lines as measured along the surface, and that this will not be the case on every kind of surface. The condition for a surface's having this important property was already established by Gauss in his famous essay on the curvature of surfaces. The condition is that what he called the 'measure of curvature', namely the reciprocal of the product of the two principal radii of curvature, should have the same magnitude throughout the whole expanse of the surface.

At the same time, Gauss showed that if the surface is bent without being stretched or shrunk in any part, this measure of curvature is not altered. Thus we can roll up a flat sheet of paper into a cylinder or cone without any alteration in the measurements of figures on the sheet, if the measurements are taken along its surface. And equally we can roll together the hemispherical closed half of a pig's bladder into the form of a spindle, without altering the measurements in this surface itself. The geometry on a plane will therefore also be the same as that in a cylindrical surface. We must only imagine in the latter case that many layers of this surface lie without limit on top of one another, like the layers of a rolled-up sheet of paper, and that with every complete circuit round the circumference of the cylinder one enters another layer, different from the one in which one was before.

These observations are necessary, in order to give you an idea of a kind of surface whose geometry is on the whole similar to that of the plane, but for which the axiom of parallels does not hold. It is a kind of curved surface which behaves in geometrical terms like the contrary of a sphere. For this reason the outstanding Italian mathematician E. Beltrami*, who investigated its properties, called it the *pseudospherical surface*. It is a saddle-shaped surface, of which only bounded pieces or strips can be displayed as a connected whole in our space. However, one can think of its being continued infinitely in all directions,

---

* *Saggio di Interpretazione della Geometria Non-Euclidea*, ["An essay at interpreting non-Euclidean geometry"], Naples, 1848. "Teoria Fondamentale degli Spazii di Curvatura Costante", ["The fundamental theory of spaces of constant curvature"], *Annali di Matematica*, 2nd series, II, 232–255.

since one can think of taking each surface portion lying at the boundary of the constructed part of the surface, pushing it back towards its middle, and then continuing it[18]. A surface portion displaced in this process must change its flexure but not its dimensions, just as one can push a sheet of paper back and forth on a cone which has arisen by rolling a plane up together. Such a sheet will everywhere adapt itself to the surface of the cone, but it must be more strongly bent nearer the tip, and it cannot be pushed on past the tip in such a way as to remain adapted to both the existing cone and its ideal continuation beyond the tip.

Pseudospherical surfaces are of constant curvature like the plane and the sphere, so that every piece of them fits exactly when laid upon any other part of the surface. Therefore, any figure constructed at one place in the surface can be transferred to any other place with a form completely congruent, and with complete equality of all dimensions lying in the surface itself. The measure of curvature laid down by Gauss, which is positive for the sphere and equal to zero for the plane, will have a constant negative value for pseudospherical surfaces, because the two principal curvatures of a saddle-shaped surface have their concavities on opposite sides[19].

It is, for example, possible to display a strip of a pseudo-spherical surface furled up as the surface of a ring. Think of a surface such as *aabb* (Figure 1), rotated about its axis of symmetry *AB*, in which case the two arcs *ab* would describe such a pseudospherical ringlike surface. The two edges of the surface, above at *aa* and below at *bb*, would bend more and more sharply outwards until the surface was perpendicular to the axis. It would end there at the rim with infinite curvature. One could also furl up half of a pseudospherical surface into a calyx-shaped champagne glass with an infinitely prolonged and tapering stem, as Figure 2. But the surface is necessarily always limited on one side by an abrupt edge beyond which a continuous extension is not immediately realisable. Only if one thinks of each individual portion of the edge being cut loose and pushed along the surface of the ring or calyx-shaped glass, can one bring such portions of the surface to locations of different flexure where they can be extended further.

In this way, the straightest lines of the pseudospherical surface can also be prolonged infinitely. They do not return upon themselves, like those of the sphere. There can instead, as on the plane, be only one

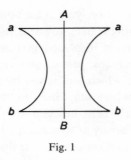

Fig. 1

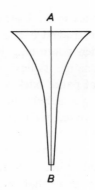

Fig. 2

single shortest line between any two given points. But the axiom of
parallels does not hold. Given a straightest line on the surface and a
point lying outside it, one can find a whole pencil of straightest lines
through the point, none of which intersects the given line even when
prolonged infinitely. They are namely all lines lying between a certain
pair of straightest lines which bound the pencil. One of these meets
the given line when infinitely prolonged in one direction, and the other
in the other direction.

Such a geometry, which drops the axiom of parallels, was moreover
completely elaborated as long ago as 1829 by the mathematician N. J.
Lobatchewsky* in Kazan, using Euclid's synthetic method[20]. It was
found that this geometrical system could be implemented in a manner
as deductive and as free of contradiction as can that of Euclid. This
geometry accords completely with that of pseudospherical surfaces,
more recently developed by Beltrami.

From all this we see that the assumption, in two-dimensional geometry,
that any figure can be moved in any direction without at all altering its
dimensions in the surface, characterises the surface concerned as being
a plane, a sphere or a pseudospherical surface. The axiom that there
exists only one shortest line between any two points separates the plane
and the pseudospherical surface from the sphere, and the axiom of
parallels divides off the plane from the pseudosphere. These three
axioms are therefore necessary and sufficient, in order to characterise

* *Prinzipien der Geometrie* ['Principles of geometry'], Kazan, 1829–30.

the surface to which Euclidean planimetric geometry refers as a plane, as opposed to all other two-dimensional spatial structures.

The distinction between geometry in a plane and on a spherical surface has been clear and intuitive for a long time. But the sense of the axiom of parallels could only be understood after Gauss had developed the concept of surfaces which can be bent without stretching, and therefore of the possibility of extending pseudospherical surfaces infinitely. We of course, being inhabitants of a three-dimensional space and endowed with sense organs for perceiving all of these dimensions, can intuitively imagine the various cases in which surface-like beings would have to develop their spatial intuition, because to this end we only have to re-strict our own intuitions to a narrower domain. It is easy to think away intuitions which one has. But to imagine with our senses intuitions for which one has never had any analogue, is very difficult. So when we go over to three-dimensional space, we are impeded in our powers of imagination by the structure of our organs and the experiences thereby acquired, which are only appropriate to the space in which we live.

However, we have also another way of dealing with geometry scien-tifically. All spatial relationships known to us are namely measurable, meaning that they can be reduced to the specification of magnitudes (lengths of lines, angles, areas, volumes). For just this reason, the prob-lems of geometry can also be solved by looking for the methods of calculation whereby the unknown spatial magnitudes have to be derived from the known ones. This occurs in *analytic geometry*, where all of the structures of space are only treated as magnitudes and specified by means of other magnitudes. Our axioms themselves also speak of spatial magnitudes [21]. The straight line is defined as the *shortest* between two points – in this one is specifying a magnitude. The axiom of parallels declares that if two straight lines in the same plane do not intersect (are parallel), the corresponding angles, or the alternate angles, made by a third line intersecting them are pairwise equal [†]. Or it is postulated, in place of this, that the sum of the angles any triangle is equal to two right angles. These too are cases of specifying magnitudes.

So one can also start from this aspect of the concept of space, according to which the situation of any point can be specified by measuring certain

---

[†] [A theorem of Euclid closely related to his postulate of parallels.]

magnitudes with respect to some spatial structure which is regarded as
fixed (coordinate system). One can then look and see what particular
specifications accrue to our space as it manifests itself in the actual
measurements to be executed, and whether there are such by which it is
distinguished from magnitudes of extension of similar manifold. This
approach was first initiated in Göttingen by Riemann*, who was sadly
lost too early to science. It has the peculiar advantage, that all of its
operations consist purely in specifying magnitudes by calculation,
whereby the danger is completely obviated that familiar, intuitive facts
might insinuate themselves as necessities of thought.

The number of measurements needed for determining the position of a
point is equal to the total number of dimensions of the space concerned.
In a line the distance from one fixed point, thus one magnitude, suffices.
In a surface one must already give the distances from two fixed points
in order to fix the position of the point. In space one must give those
from three, so that we need, as on the earth, longitude, latitude and
height above sea-level, or, as is usual in analytic geometry, the distances
from three coordinate planes. Riemann terms a system of differences
in which the individual can be specified by *n* measurements, a *n-fold
extended manifold*, or a *manifold of n dimensions*. Thus the space known
to us, in which we live, is in these terms a threefold extended manifold
of points. A surface is a twofold one, a line a onefold one, and time
equally a onefold one. The colour system too consists of a threefold
manifold, inasmuch as every colour, according to Thomas Young's and
Maxwell's** investigations, can be described as a mixture of three basic
colours, of each of which a definite amount is to be used. Using the
coloured spinning top, one can actually carry out such mixtures and
measurements.

We could equally regard the realm of simple tones*** as a manifold
of two dimensions, if we just distinguished them by pitch and intensity
and disregarded differences of timbre[23]. This generalisation of the concept is very suitable for bringing out what distinguishes space from other

* 'Über die Hypothesen, welche der Geometrie zu Grunde liegen' ['On the hypotheses
underlying geometry'], inaugural dissertation of 10 June 1854. Published in vol. XIII of
the *Abhandlungen der königlichen Gesellschaft zu Göttingen*[22].
** See [Helmholtz'] *Vorträge und Reden*, vol. I, p. 307.
*** *Vorträge und Reden*, vol. I, p. 141.

manifolds of three dimensions [24]. In space, as you all know from every-day experience, we can compare the distance between two points lying one above the other with the horizontal distance between two points on the floor, because we can apply one measuring rod alternately to the one pair and to the other. But we cannot compare the distance between two tones of equal pitch and different intensity with that between two tones of equal intensity and different pitch.

By considerations of this kind, Riemann showed that the essential basis of any geometry is the expression giving the separation between two points for any arbitrary direction of separation, and to be exact that in the first place between two infinitesimally separated points. From analytic geometry, he obtained the most general form* which this ex-pression takes for a completely arbitrary choice of the kind of measure-ments yielding the place of each point. He then showed that the kind of freedom of motion without distortion which characterises bodies in our space, can only exist if certain magnitudes** emerging from the calcu-lation have the same value everywhere. In respect of the relationships in surfaces, these magnitudes reduce to the Gaussian measure of curvature. For this very reason, when these calculated magnitudes have the same value in all directions from a given location, Riemann terms them the measure of curvature, at that location, of the space concerned.

To prevent misunderstandings, I just wish also to emphasize here that this so-called measure of curvature of space is a calculated magnitude obtained in a purely analytic way, and that its introduction in no way rests upon insinuating relationships which would only have a sense as ones intuited by the senses. The name, as a brief expression for denoting a complicated relationship, has simply been adopted from the one case where the denoted magnitude corresponds to something intuited by the senses [25].

If moreover this measure of curvature of space has everywhere the value zero, such a space corresponds everywhere to the axioms of Euclid. In this case we can term it a *flat* space, in contrast with other

---

* Namely, for the square of the distance between two infinitely close points a homoge-neous second degree function of the differentials of their coordinates.
** It is an algebraic expression which is constructed out of the coefficients of the individual terms in the expression for the square of the separation between two neighbouring points and their derivatives.

analytically constructible spaces which one might term curved, because their measure of curvature has a value different from zero. At the same time, the analytic geometry of spaces of the latter kind can be worked out as completely and self-consistently, as the customary geometry of our in fact existing[26] flat space. For a positive measure of curvature we obtain *spherical* space, in which straightest lines return upon themselves and in which there are no parallel lines. Such a space would, like the surface of a sphere, be unbounded but not infinitely large. On the other hand, a constant negative measure of curvature gives *pseudospherical* space, in which straightest lines go on to infinity, and where in any flattest surface one can find a pencil of straightest lines through any point which do not intersect some other given straightest line of that surface[27].

This latter situation has been made intuitively accessible by Beltrami*, through his showing how one can form an image of the points, lines and surfaces of a three-dimensional pseudospherical space in the interior of a sphere of Euclidean space, and in such a way that every straightest line of the pseudospherical space is represented in the sphere by a straight line, and every flattest surface of the former by a plane in the latter. The surface itself of the sphere corresponds here to the infinitely distant points of the pseudospherical space. In their image in the sphere, the various parts of that space become progressively smaller as they approach the surface of the sphere, and indeed more strongly in the direction of its radii than in directions perpendicular to them. Straight lines in the sphere which intersect only outside its surface, correspond to straightest lines of the pseudospherical space which never intersect.

It thus emerged that space, considered as a domain of measurable magnitudes, by no means corresponds to the most general concept of a three-dimensional manifold. Instead, it contains further particular specifications, which are occasioned by the completely free mobility without distortion of fixed bodies to every place and with every possible change of direction. And occasioned too by the particular value of the measure of curvature, which is to be set equal to zero for the in fact existing space, or whose value is at least not noticeably distinct from zero. This

* 'Teoria Fondamentale degli Spazii di Curvatura Costante' (*op. cit.*).

last stipulation is given in the axioms of straight lines and of parallels.

Riemann entered this new field by starting from the most general basic questions of analytic geometry. I myself had meanwhile arrived at similar considerations, partly through investigations into portraying the colour system spatially, thus through comparing one threefold extended manifold with another, and partly through investigations into the origin of our visual estimation for measurements in the visual field. Riemann started out from the above mentioned algebraic expression, which describes in the most general form the separation between two infinitely close points, as his basic assumption, and he derived from it the theorem on the mobility of fixed spatial structures, whereas I started out from the observational fact[28], that in our space the motion of fixed spatial structures is possible with that degree of freedom with which we are acquainted, and I derived from this fact the necessity of the algebraic expression which Riemann set down as an axiom. The following were the assumptions on which I had to base the calculation[29].

It must *firstly* be presupposed – in order to make treatment by calculation possible at all – that the situation of any point A can be specified by measuring certain spatial magnitudes, whether lines or angles between lines or angles between surfaces or whatever else, with respect to certain spatial structures which are regarded as unchangeable and fixed. As is well known, one terms the measurements needed for specifying the situation of the point A its *coordinates*. The total number of coordinates needed in general for specifying completely the situation of any point whatever, specifies the total number of dimensions of the space concerned. It is further presupposed that the spatial magnitudes used as coordinates change continuously as the point A moves.

One must *secondly* give such a definition of a fixed body, or fixed point system, as is needed to enable comparison of spatial magnitudes by congruence. As we should not yet presuppose here any special methods of measuring[30] spatial magnitudes, the definition of a fixed body can now only be given by the following characteristic: there must exist, between the coordinates of any two points belonging to a fixed body, an equation which expresses an unchanged spatial relationship between the two points (which finally turns out to be their separation) for any motion of the body, and one which is the same for congruent point pairs, while point pairs are congruent, if they can successively coincide with the same point pair fixed in space[31].

This definition has extremely far-reaching implications, although its formulation seems so incompletely specific, because with an increasing number of points the total number of equations grow much faster than the number of the coordinates (of these points) that they determine. Five points, A, B, C, D and E, give ten different point pairs:

$$AB, \ AC, \ AD, \ AE,$$
$$BC, \ BD, \ BE,$$
$$CD, \ CE,$$
$$DE,$$

and so ten equations, which in three-dimensional space contain fifteen variable coordinates. Of these, however, six must remain freely disposable if the system of five points is to be freely movable and rotatable. Thus the ten equations are to determine only nine coordinates, as dependent upon the six variable ones. With six points we get fifteen equations for twelve variable magnitudes, with seven points twenty-one equations for fifteen magnitudes, and so on.

However, from n mutually independent equations we can determine n magnitudes appearing in them. If we have more than n equations, the surplus ones must themselves be derivable from the first n of them. From this it follows that the equations which exist between the coordinates of each point pair of a fixed body must be of a particular kind, namely such that if they are satisfied in three-dimensional space for nine of the point pairs formed from any five points, the equation for the tenth pair must follow from them identically. It is in virtue of this circumstance, that the stated assumption for the definition of fixity is indeed sufficient in order to specify the kind of equations which exist between the coordinates of two fixedly interconnected points.

It resulted *thirdly* that the calculation had to be based on the fact of yet one more particular peculiarity of the motion of fixed bodies, a peculiarity so familiar to us, that without this investigation we might never have chanced to regard it as something that might also not be so. If we namely hold fixed two points of a fixed body in our three-dimensional space, it can then only rotate about the straight line connecting them as axis of rotation. If we rotate it once right round, it comes back exactly to the situation in which it was initially. Now it has to be specifically stated, that rotation of any fixed body without reversal always brings it back to its initial situation[32]. There might be a geometry where this was not so. This is to be seen most simply for the geometry of the plane. If one thinks of every plane figure increasing its linear dimensions under any rotation in proportion to the angle of rotation, then after a whole rotation through 360 degrees the figure would no longer be congruent with its initial state. Yet of course any other figure which was congruent with it in its initial situation could also be made congruent with it in its second situation, if the second figure too were rotated through 360 degrees. A self-consistent system of geometry, not falling within Riemann's scheme, would be possible under this assumption too.

I have shown, on the other hand, that the enumerated three assumptions are together sufficient in order to establish the starting point assumed by Riemann in the investigation. With it are also established all further results of his work which concern the distinction of the various spaces according to measure of curvature.

One might yet ask whether the laws of motion, and of its dependence on motive forces, can also be carried over without contradiction to spherical or pseudospherical space. This investigation has been carried out by Herr Lipschitz* in Bonn. One can actually carry over the comprehensive expression of all laws of dynamics, namely Hamilton's Principle[33], directly to spaces whose measures of curvature are not equal to zero. Thus the deviating systems of geometry are not subject to contradiction in this respect either.

We shall now have to go on and ask for the origin of these particular specifications [which characterise our space as a flat space]†, since it has emerged that they are not included in the general concept of an extended three-dimensional magnitude and of the free mobility of the bounded structures contained in it. They are not *necessities of thought*[34], deriving from the concept of such a manifold and its measurability, or from the most general concept of a fixed structure contained in it and of the freest mobility of the latter.

We now wish to investigate the contrary assumption that may be

---

* 'Untersuchungen über die ganzen homogenen Funktionen von *n* Differentialen' ['Investigations on total homogeneous functions of *n* differentials'], *Borchardts Journal für Mathematik* **70**, 71 and **72**, 1. 'Untersuchung eines Problems der Variationsrechnung' ['Investigation of a problem in the calculus of variations'], *ibid.*, **74**.
† [Hertz and Schlick appear to have omitted this clause by mistake.]

made about their origin, namely the question of whether their origin is *empirical*, whether they are to be derived or made evident from facts of experience, or correspondingly to be tested and perhaps even refuted by these[35]. This latter eventuality would then include too, that we should be able to imagine series of observable facts of experience by which another value of the measure of curvature would be indicated, than that of Euclid's flat space. If indeed spaces of another kind are imaginable in the sense stated, this will also refute the claim that the axioms of geometry are in Kant's sense necessary consequences of a transcendental form, given *a priori*, of our intuitions[36].

As noted above, the distinction between Euclidean, spherical and pseudospherical geometry rests upon the value of a certain constant, which Riemann terms the measure of curvature of the space concerned, and whose value must be equal to zero if Euclid's axioms hold. If it is not equal to zero, triangles of large area will necessarily have a different sum of angles from small ones, the former having a larger sum in spherical and a smaller one in pseudospherical space. Moreover, geometrical similarity between larger and smaller bodies or figures is only possible in Euclidean space. All systems of practically executed geometrical measurements in which the three angles of large rectilinear triangles have been measured individually, and so in particular all systems of astronomical measurements which yield zero for the parallax[37] of immeasurably distant fixed stars (in pseudospherical space even the infinitely distant points must have positive parallax), confirm the axiom of parallels empirically[38]. They show that in our space and using our methods of measurement, the measure of curvature of space appears to be indistinguishable from zero. One must of course raise, with Riemann, the question of whether this perhaps might not be different if instead of our limited base lines, of which the greatest is the major axis of the earth's orbit, we could use greater base lines.

But we should not forget here, that all geometrical measurements rest ultimately on the principle of congruence[39]. We measure separations between points by moving a pair of dividers or measuring rod or measuring chain up to them. We measure angles by bringing a protractor or theodolite up to the vertex of the angle. We determine straight lines, additionally, by the path of light rays too, which in our experience is rectilinear; but it could also be carried over equally to spaces of other measures of curvature that light, as long as it stays in an unchanging refracting medium, is propagated along shortest lines. So all our geometrical measurements rest upon the presupposition that the measuring instruments which we take to be fixed, actually[40] are bodies of unchanging form. Or that they at least undergo no kinds of distortion other than those which we know, such as those of temperature change, or the small extensions which ensue from the different effect of gravity in a changed location.

When we measure, we are only doing, with the best and most reliable auxiliary means known to us, the same thing as what we otherwise ordinarily ascertain through observation by visual estimation and touch, or through pacing something off. In the latter cases, our own body with its organs is the measuring instrument which we carry around in space. At different moments our hand or our legs are our dividers, or our eye turning in all directions is the theodolite with which we measure arcs or plane angles in the visual field.

Thus any comparative estimation of magnitudes, or measurement of spatial relationships, starts from a presupposition about the physical behaviour of certain bodies, whether of our own body or of applied measuring instruments. This presupposition may incidentally have the highest degree of probability[41] and be in the best agreement with all physical circumstances otherwise known to us, but it still goes beyond the domain of pure spatial intuitions.

One indeed can instance a certain behaviour[42] of the bodies that appear fixed to us, such that measurements in Euclidean space would turn out as if they had been performed in pseudospherical or spherical space. To see this[43], I will point out first that if the whole of the linear dimensions of the bodies surrounding us, and with them those of our own body, were all decreased in the same proportion – say by half, or say they were all increased to double their size – we would be quite unable to notice such a change with our means of spatial intuition. The same would moreover also be the case if the extension or contraction were different in different directions, provided that our own body changed in the same way, and provided further that a rotating body adopted, without undergoing or exercising mechanical resistance, at every moment the degree of extension of its various dimensions that corresponded to its momentary situation[44].

Think of the image of the world in a convex mirror. The familiar silvered globes that are commonly set up in gardens display the essential phenomena of such an image, if indeed distorted by some optical ir-regularities. A well-made convex mirror of not too great an aperture displays, in a definite situation and at a definite distance behind its sur-face, the apparently corporeal image of any object lying in front of it. But the images of the distant horizon and of the sun in the sky lie behind the mirror at a limited distance, equal to its focal length. The images of all other objects lying in front of it are contained between those images and its surface, but such that the images are progressively more dimin-ished and flattened with increasing distance of their objects from the mirror. The flattening, in other words the diminution in the dimension of depth, is relatively more pronounced than the diminution in the areal dimensions. Nonetheless, every straight line of the outer world is por-trayed by a straight line in the image, and every plane by a plane. The image of a man measuring, with a measuring rod, a straight line stretch-ing away from the mirror, would progressively shrivel up as its original moved away. But the man in the image would count up, with his equally shrivelled measuring rod, exactly the same number of centimeters as the man in actuality. In general, all geometrical measurements of lines or angles carried out with the regularly changing mirror images of the actual instruments, would yield exactly the same results as those in the outer world. All congruences would match in the images, if the bodies concerned were actually laid against each other, just as in the outer world. All lines of sight in the outer world would be replaced by straight lines of sight in the mirror.

In short, I do not see how the men in the mirror could bring it out that their bodies were not fixed bodies and their experiences [not] good examples of the correctness of Euclid's axioms. But if they could look out into our world, as we look into theirs, without being able to cross the boundary, then they would have to pronounce our world to be the image of a convex mirror, and speak of us just as we of them. And as far as I can see, if the men of the two worlds could converse together, then neither would be able to convince the other that he had the true and the other the distorted situation. I cannot even recognise that such a question has at all any sense[45], as long as we introduce no mechanical considerations[46].

Now Beltrami's mapping[47] of pseudospherical space onto a whole sphere of Euclidean space is of a quite similar kind, except that the background surface is not a plane, as with the convex mirror, but the surface of a sphere, and that the proportion in which the images contract, as they approach the surface of the sphere, has a different mathematical expression. Suppose one therefore thinks conversely of bodies which move in the sphere in whose interior Euclid's axioms hold, and which always contract, like the images in the convex mirror, when moving away from the centre, and which moreover contract in such a way that their images constructed in pseudospherical space preserve unchanged dimensions. Then observers whose own bodies regularly underwent this change would obtain results from geometrical measurements, made as they could make them, as if they themselves lived in pseudospherical space.

Starting from here we can go yet one step further. We can deduce how the objects of a pseudospherical world would appear to an observer, whose visual estimation and spatial experiences had, exactly as ours, been developed in flat space, if he could enter such a world. Such an observer would continue to see the lines of light rays, or the lines of sight of his eyes, as straight lines like those existing in flat space, and like they actually are in the spherical image of pseudospherical space. The visual image of the objects in pseudospherical space would therefore give him the same impression as if he were at the centre of Beltrami's spherical image[48]. He would believe he could see all round himself the most distant objects of this space at a finite distance*, let us for example assume at a distance of 100 feet. But if he approached these distant objects they would expand in front of him, and indeed more in depth than in area, while behind him they would contract. He would discern that he had judged by visual estimation falsely.

Suppose he saw two straight lines which, in his estimation, stretched away parallel for this whole distance of 100 feet, to where the world seemed to end for him. On pursuing them he would then discern that, together with this expansion of the objects he approached, these lines would spread apart the more he advanced along them. Behind him, on the other hand, the distance between them would seem to dwindle, so that as he advanced they would seem to be progressively more divergent

---

* The negative inverse square of this distance would be the measure of curvature of the pseudospherical space.

and separated from each other. But two straight lines which seemed, from the first viewpoint, to converge to one and the same point in the background at a distance of 100 feet, would always do this however far he went, and he would never reach their point of intersection.

Now we can obtain quite similar pictures of our actual world, if we put before our eyes a large convex lens of corresponding negative focal length[†], or indeed only two convex eye glasses, which would have to be ground slightly prismatically as if they were parts of a larger, connected lens. These, just like the convex mirror mentioned above, show us distant objects brought up closer, the most distant ones up to the distance of the focal point of the lens. If we move with such a lens before our eyes, there take place quite similar expansions of the objects we approach to those I have described for pseudospherical space. If, moreover, someone puts such a lens before his eyes, and not even a lens of focal length 100 feet, but a much stronger one of focal length only 60 inches, then perhaps for the first moment he notices that he sees objects brought up closer. But after a little going back and forth the illusion fades, and he judges distances correctly despite the false images. We all have grounds to surmise that it would soon enough be the same for us in pseudospherical space, as it is after just a few hours for someone who begins to wear glasses. In short, pseudospherical space would seem to us to be relatively not very strange at all. Only at first would we find ourselves subject to illusions, in measuring the magnitude and distance of more distant objects by their visual impression.

A three-dimensional spherical space would be accompanied by the opposite illusions, if we entered it with the visual estimation acquired in Euclidean space. We would take more distant objects to be further off and larger than they were. On approaching them, we would find ourselves reaching them more quickly than we should assume from their visual image. However, we would also see before us objects which we could only fixate with diverging lines of sight: this would be the case with all those more distant from us than a quadrant of a great circle. This kind of view would hardly strike us as very unusual, since we can produce it for earthly objects too, by putting before one eye a weakly prismatic glass whose thicker side is turned towards the nose. Then too

---

[†] [i.e. a concave lens.]

we have to set our eyes divergently in order to fixate distant objects. It produces in the eyes a certain feeling of unaccustomed strain, but does not notably change the visual features of objects seen thus, whereas the strangest visual feature of the spherical world would consist of the back of our own head, upon which all of our lines of sight would reconverge, to the extent that they could pass freely between other objects, and which would have to fill up the furthest background of the whole perspective image.

At the same time, the following is of course also to be noted. Just as a small flat elastic plate, say a small flat rubber disk, can only be accommodated to a weakly curved spherical surface with a relative contraction of its edge and stretching of its middle, so also our own body, which has grown up in Euclidean flat space, could not go over into a curved space without undergoing similar stretchings and compressions of its parts. Naturally, the interconnexion of these parts could only be maintained as long as their elasticity allowed yielding without tearing and breaking. The kind of stretching would have to be the same as if we thought of a small body at the centre of Beltrami's sphere, and then transposed from it to its pseudospherical or spherical image. For such a transposition to appear possible, it must always be presupposed that the body transposed is sufficiently elastic and small, in comparison with the real or imaginary radius of curvature of the curved space into which it is to be transposed.

This will suffice to show how one can deduce from the known laws of our sense perceptions, continuing in the way begun here, the series of sense impressions which a spherical or pseudospherical world would give us if it existed. In this respect too we nowhere meet an impossibility or deductive fault, just as little as in the calculative treatment of the metrical relationships. We can just as well depict the view in all directions, in a pseudospherical world, as we can develop its concept. For this reason, we also cannot admit that the axioms of our geometry are based upon the given form of our faculty of intuition, or are connected with such a form in any way[49].

It is otherwise with the three dimensions of space. All our means of intuition by the senses only stretch to a space of three dimensions, and the fourth dimension would not be a mere modification of what exists, but something completely new. Thus if only on account of our bodily makeup, we find ourselves absolutely unable to imagine a way of intuitively conceiving a fourth dimension[50].

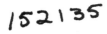

I wish finally to stress further, that the axioms of geometry are certainly not propositions belonging to the pure theory of space alone[51]. As I have already mentioned, they speak of magnitudes. One can only talk of magnitudes if one knows and intends some procedure, whereby one can compare these magnitudes, split them up into parts and measure them. Thus all spatial measurement, and therefore in general all magnitude concepts applied to space, presuppose the possibility of the motion of spatial structures whose form and magnitude one may take to be unchanging despite the motion. In geometry, such spatial forms are indeed by custom referred to only as geometrical bodies, surfaces, angles and lines, because one abstracts from all other distinctions of a physical and chemical kind manifested by natural bodies. But one retains nonetheless one of their physical properties, namely fixity. Now we have no criterion for the fixity of bodies and spatial structures other than that when applied to one another at any time, in any place and after any rotation, they always show again the same congruences as before. But we certainly cannot decide in a purely geometrical way, without bringing in mechanical considerations[52], whether the bodies applied to each other have not themselves both changed in the same manner.

If we ever found it useful for some purpose, we could, in a completely logical way, regard the space in which we live as the apparent space behind a convex mirror with its shortened and contracted background. Or we could regard a bounded sphere of our space, beyond whose bounds we perceive nothing more, as infinite pseudospherical space. We would then only have to ascribe the corresponding stretchings and shortenings to the bodies which appear to us to be fixed, and equally to our own body at the same time. We would of course at the same time have to change our system of mechanical principles completely. For even the proposition, that every moving point upon which no force acts continues to move in a straight line with unchanged velocity, no longer applies to the image of the world in the convex mirror. The path would indeed still be straight, but the velocity dependent upon place[53].

Thus the axioms of geometry certainly do not speak of spatial relationships alone, but also, at the same time, of the mechanical behaviour of our most fixed bodies during motions[54]. One could admittedly also take the concept of fixed geometrical spatial structures to be a transcendental[55] concept, which is formed independent of actual experiences and to which these need not necessarily correspond, as in fact our natural bodies are already not even in wholly pure and undistorted correspondence to those concepts which we have abstracted from them by way of induction. By adopting such a concept of fixity, conceived only as an ideal, a strict Kantian certainly could then regard the axioms of geometry as propositions given *a priori* through transcendental intuition, ones which could be neither confirmed nor refuted by any experience, because one would have to decide according to them alone whether any particular natural bodies were to be regarded as fixed bodies. But we would then have to maintain that according to this conception, the axioms of geometry would certainly not be synthetic propositions in Kant's sense. For they would then only assert something which followed analytically from the concept of the fixed geometrical structures necessary for measurement, since only structures satisfying those axioms could be acknowledged to be fixed ones.

But suppose we further add propositions concerning the mechanical properties of natural bodies to the axioms of geometry, if only the law of inertia, or the proposition that the mechanical and physical properties of bodies cannot, under otherwise constant influences, depend upon the place where they are. Then such a system of propositions is given an actual content, which can be confirmed or refuted by experience, but which for just that reason can also be obtained by experience[56].

Incidentally, I naturally do not intend to maintain that humankind only obtained intuitions of space, corresponding to Euclid's axioms, by carefully executed systems of exact geometrical measurements. A series of everyday experiences, and especially the intuition of the geometrical similarity of larger and smaller bodies – which is only possible in flat space – must rather have led to the rejection as impossible of every geometrical intuition opposed to this fact. For this no knowledge was needed of the conceptual connexion between the observed fact of geometrical similarity and the axioms, but only an intuitive acquaintance with the typical behaviour of spatial relationships, obtained by observing them frequently and exactly, namely the kind of intuition of objects to be portrayed which the artist has, and by means of which he decides, with assurance and refinement, whether or not an attempted new combination corresponds to the nature of the object to be portrayed. We indeed know of no other name in our language to refer to this but 'intuition', but it is an empirical acquaintance, obtained by the accumulation and reinforcement in our memory of impressions which recur in the same manner. It is no transcendental form of intuition given before all experience.

That this sort of intuition of a typical lawlike occurrence, gained empirically and not yet developed to the clarity of a definitely expressed concept, has often enough imposed itself upon metaphysicians as a proposition given *a priori*, is a matter which I need not discuss further here.

NOTES AND COMMENTS

[1] In logic one understands by 'deduction' the derivation of a judgment from more comprehensive judgments, i.e. from ones having more general validity. It is the only procedure giving a completely *rigorous* foundation for a truth by means of other truths. The most usual form of deduction is the syllogism (compare note IV.43)[†]. Deduction stands in opposition to logical "induction", which endeavours to infer more generally valid truths from particular and individual ones, and which, to this end, must submit to the "troublesome and tedious task" of gathering empirical facts, of which Helmholtz speaks in this same sentence. Inductive inference lacks absolute certainty, for inasmuch as it extends, by generalization, a proposition extracted from a number of individual cases to cases that have not yet been observed[††], it goes beyond the actual content of the presuppositions. Thus propositions obtained by induction can only claim to hold with *probability* (though with frequently an extremely high one).

[2] In this question Kant summarizes the problem of his *Kritik der reinen Vernunft* [Critique of Pure Reason].

By a 'synthetic judgment' he understands a proposition which attributes to an object a predicate that does not belong to the object already in virtue of its definition. As against this, an 'analytic judgment' only asserts of a subject something that is already contained in this subject by definition. If, for instance, by a 'body' I always understand something extended, then the proposition 'all bodies are extended' is evidently analytic, whereas we would have a synthetic judgment if we said 'all bodies are subject on the surface of the earth to an acceleration of about 981 cm/sec^2'. The first judgment follows simply from the definition, from the concept of a body. The second cannot be derived from this, but only experience can teach us whether, and with what acceleration, the objects characterised by the concept 'body' fall towards the earth.

A judgment whose validity is assured wholly independent of any experience is termed by Kant '*a priori*' ; if its validity rests only upon experience, it is called '*a posteriori*' . It follows from what has been said that every analytic judgment is *a priori*, for indeed one needs no experience to perceive its validity, but only an analysis of the concept of its

---

[†] [In modern logic the syllogism no longer has a position of any prominence; its significance is now historical as the first fragment of logic to have been analyzed with full logical rigour.]

[††] [This is not the only kind of inductive inference: see R. Carnap, *Logical Foundations of Probability*, 2nd ed., 1962, §44B. The whole subject is still controversial.]

subject, and this concept is completely given by its definition. It also follows that most synthetic judgments, both in everyday life and in science, are *a posteriori*, because they express the results of experience.

Are there also synthetic *a priori* judgments, thus in other words propositions which assert *more* about an object than already exists in its concept, yet without drawing this 'more' from experience? One easily recognizes the enormous importance of the question. For only in synthetic judgments is there a real advance of knowledge; only they extend our knowledge, while analytic ones merely elucidate what we have put into our concepts by definition. But if all synthetic judgments are only *a posteriori*, then no advance of knowledge occurs with absolutely certain validity, since a validity resting upon items of experience reaches no further than these very items themselves and may be overthrown by new observations.

We are only certain of the universal, necessary validity of a proposition if it is valid *a priori*. Necessity, and validity without exception, are therefore the characteristic by which the *a priori* is recognized. According to Kant, this characteristic attaches to the axioms of geometry, and he also, as noted by Helmholtz in the text, does not doubt that they are synthetic. Thus he believed that synthetic *a priori* judgments, this highest form of knowledge, indeed exist, and the only question for him was: how are they possible? How can I with certainty assert of an object something which is neither deduced from its definition nor drawn from experience? Helmholtz' investigation, as will emerge, examines the question as to whether the axioms of geometry actually should, as Kant presupposed, be regarded as synthetic *a priori* judgments.

[3] In taking space to be a 'form of intuition' , and hence a lawlike feature characteristic of our intuiting consciousness, Kant precisely wanted to explain the possibility of synthetic *a priori* judgments about space. Everything that we can perceive and intuitively imagine must necessarily be subject to the laws governing the manner in which we intuit. Thus these laws, according to Kant, must be valid *a priori* for everything that can be experienced.

[4] "...schema devoid of any content" – this will later need comment in greater detail (compare note IV.33).

[5] The definition of geometry as 'the theory of space' suffices at this point, although it raises several questions – especially whether one can state more precisely what is to be understood here by 'space' . It could, for instance, be that the mathematician, the physicist and the psychologist do not at all mean the same thing when speaking of 'space' . Helmholtz' investigation will also lead of itself to this question.

[6] This list of 'axioms' is clearly to be taken as only a collection of examples, intended to indicate what is involved. Helmholtz clearly does not want to say here that precisely these propositions, and in exactly this formulation, must appear as axioms in a self-contained system of geometry.

[7] The great successes of the method of logical inference in mathematics explain the wish to give a proof for everything, and at the same time the belief in the success of such an undertaking (an endeavour to which, in the field of philosophy, the systems of Fichte and Reinhold owe their origin; both tried to derive a whole system from *one* single proposition). However, the simple consideration that every inference needs premises shows the

impossibility of proving everything, and the necessity of making presuppositions. Such starting points of logical thought are the axioms, whose origin is precisely to be investigated here.

[8] The prototype of the idealist philosopher is Plato. He explains the self-evidence of the axioms of geometry by the hypothesis of *'anamnesis'* or 'recollection'. In earlier stages of its existence, before its earthly birth, the human soul supposedly became acquainted with the truths of mathematics through immediate contemplation, and it now knows about them without earthly experience, by mere recollection.

It perhaps deserves to be mentioned that in this Platonic doctrine the explanation is, properly speaking, still based on experience, even though not on that of the senses which belongs to earthly life, but on an entirely different kind of experience during a mythical pre-existence.

[9] This of course does not yet provide a rigorous proof of the fruitlessness of the attempts. The impossibility of deriving a proposition from certain axioms (i.e. the "independence" of this proposition from those axioms, which requires its being placed with the other axioms as a new one) is in general proved by dropping the proposition in question – or replacing it with another – while retaining all the other axioms, and testing the system of axioms which results for freedom from contradiction. Thus, for instance, the independence of the axiom of parallels from the other Euclidean axioms follows from the fact, that by giving it up we variously arrive at the geometrical system of Bolyai and Lobatchewsky or at that of Riemann. Now one can prove that these two non-Euclidean geometries are free from internal contradiction, *provided* that Euclidean geometry is. The freedom of the latter from contradiction (which was probably never in doubt) can be proved if one presupposes that the structure of number theory – arithmetic – is free from contradiction. To date nobody seems to have succeeded in rigorously demonstrating the correctness of this presupposition[†].

[10] The Greek mathematician Euclid compiled, around 300 B.C., the mathematical knowledge of his time in a text-book of such excellence that it occasionally (e.g. in English schools) serves as the basis for teaching geometry even to this day.

[11] In talking, as Helmholtz does here, of 'movement' as something that can in fact be experienced, as an actual process, one must by 'geometrical structures' understand here rigid corporeal models, and not purely mathematical lines or surfaces as mere structures of abstraction. There arises the important question of to what extent a distinction can be made at all between 'geometrical' and 'corporeal' structures, thus the problem of the relation of geometry to physics, on which Helmholtz, towards the end of the lecture, quite specifically adopts a position.

––––––––

[†] A number of proofs have been given for the consistency of the arithmetic of the natural numbers, starting with G. Gentzen, 'Die Wiederspruchsfreiheit der reinen Zahlentheorie' ['The freedom from contradiction of pure number theory'], *Math. Annalen*, CXII (1936), 493–565. However, all of these proofs have the methodological drawback that they can be formulated only in a mathematical form of language (e.g. mathematical and even transfinite induction are used). There are moreover theoretical reasons for supposing that this drawback is insuperable. See Elliott Mendelson, *Introduction to Mathematical Logic*, Van Nostrand, Princeton, 1964, Appendix and pp. 148–9.

[12] It should be noted that the conclusion drawn by Helmholtz in the last sentence is only admissible, if it is true that every presupposition must either have a logical foundation or have originated in experience, and thus cannot have come from a third source.

[13] 'Analytic' geometry reduces all proofs to calculations through denoting each point of space by three numbers (coordinates), for instance by giving its distances from three mutually perpendicular planes (compare p. 13 of the text). In this way it becomes possible to treat, in a purely calculative manner, the relations between spatial structures as numerical relations (namely between the coordinates of their points).

[14] 'purely logical operation' , i.e. deduction (compare note 1).

[15] As it is appropriate to restrict the term 'to think' to the logical operations of judgment and inference, the first of the two formulations, 'to imagine', is doubtless to be preferred. The sense of the explanation given by Helmholtz in this sentence can hardly be misunderstood in its simplicity. 'Imaginable' means everything that is intuitively reproducible in the sense of psychology. Although a certain subjectivity or relativity is inherent in this definition, due to individual differences of imaginative capacity, it should not affect the *basic* meaning of this delimitation of the concept 'imaginable' (compare note IV.42).

[16] The concept of the 'straightest' path also plays a decisive role in the mechanics of Heinrich Hertz.

[17] Thus J. Wallis (1614–1703) substituted for the Euclidean axiom of parallels the proposition: for any figure there exists a similar one of arbitrary magnitude. See Bonola-Liebmann, *Die nichteuclidische Geometrie* ['Non-Euclidean geometry'], Leipzig, 1908, pp. 14f.

[18] D. Hilbert (*Grundlagen der Geometrie* ['Foundations of geometry'], 3rd ed., Leipzig and Berlin, 1909, p. 251) has proved that "there does not exist an analytic surface of constant negative curvature which is without singularities and everywhere regular. Thus in particular, the question of whether one can realize in Beltrami's manner the whole of the Lobatchewskian plane by means of a regular analytic surface in space, must also be answered in the negative." But this mathematical result, to which Riehl also refers (*Helmholtz in seinem Verhältnis zu Kant* ['Helmholtz in his relation to Kant'], 1904, p. 37), is not of fundamental significance for the basic epistemological thought behind Helmholtz' lecture.

[19] ...and the radii of curvature are therefore taken to have opposite signs.

[20] 'Euclid's synthetic method' is the same as what Helmholtz on p. 4 called the 'intuitive method'.

[21] Namely the axioms of our ordinary metric geometry. One can also think of a geometric system containing no concepts of magnitude at all. Such is *analysis situs*, which deals exclusively with those properties of spatial structures which contain no *metric* concepts. For this reason Poincaré terms *analysis situs* the true "*qualitative* geometry" (*Der Wert der Wissenschaft* ['The value of science'], 2nd ed., 1910, p. 49).

[22] Riemann's dissertation has recently been published with detailed comments by H. Weyl, Berlin, 1919, 2nd ed., 1921.

23 Thus in the manifold of colours, the three amounts of the basic colours needed for mixing a specific colour would have to be regarded as the three 'coordinates' of the latter (see notes II.8, III.29); in the manifold of tones, the intensity and pitch of each tone would be its two coordinates.

24 The distinguishing characteristic stressed by Helmholtz in the following is not sufficient, though indeed necessary, to specify what is proper to 'space' as against other three-dimensional manifolds. To what extent can one at all speak of a 'definition' of space? What spatial extension is in the psychological-intuitive sense, can only become known through perceptual experiences; it is just as little definable, as it is possible to give a person born blind an idea of what 'red' is. Only those properties of space are to be regarded as definable which are accessible to mathematical analysis. But these are the sole properties that physics has to make use of, and from this one can draw important conclusions.

25 One can hardly draw attention emphatically enough to what Helmholtz is stressing here: the expression 'measure of curvature' in its application to space should only ever be understood in a metaphorical sense, it is not intended to denote anything that we can in any way intuitively perceive or imagine as a curvature. A typical example of this wide-spread misunderstanding is found in the polemic *Kant und Helmholtz* ['Kant and Helm-holtz'] by Albrecht Krause, Lahr, 1878, where it is said on p. 84: "Lines, surfaces, axes of bodies in space have an orientation and thus also a measure of curvature, but space as such has no orientation, precisely because everything oriented is in space, and it there-fore has *no* measure of curvature; but this is something other than having a measure of curvature *equal to zero*."
Another misunderstanding, frequently met, is connected with this. Since we are only able to imagine a curved surface in a three-dimensional space, it is namely inferred that a 'curved' three-dimensional manifold presupposes the existence of a four-dimensional one. But this inference from analogy is false. Were it correct, then 'flat' Euclidean space would also have to be imaginable only as embedded in a four-dimensional space: it is namely evident that we also cannot imagine a flat surface otherwise than as in space, since certainly for our imagination it always remains possible to leave the flat surface, just as a curved one, by stepping out on either side into space. The layman too easily forgets that 'curvature' in the Gaussian sense is a wholly *internal* property of the surface, and that a curved surface is just as much a merely two-dimensional structure as a flat one.
Helmholtz says correctly that in the case of a surface an intuition of the senses corre-sponds to the 'curvature'. But not every surface which is curved for intuition is thus also in the mathematical sense. Notably, the surfaces of e.g. a cone or a cylinder have the Gaussian measure of curvature zero.

26 Helmholtz speaks of our "in fact existing" space; he never doubted that physical space is not merely a partially arbitrary mental construction, but something actual, whose prop-erties can be ascertained by observation. In what sense this presupposition is founded and justified we shall see from Helmholtz' own statements. Here he describes the in fact existing space as 'flat'; on p. 18 he says further that besides flatness a tiny measure of curvature, not noticeably different from zero, is also compatible with the data of experience.

27 Further types of space again are elliptical space and Klein-Clifford space; compare note II.30.

[28] Helmholtz terms it an 'observational fact' that motion of fixed spatial structures, i.e. change of place of rigid bodies, is possible. To what extent this should actually count as an observational fact will be discussed later on (note 40).

[29] The mathematical developments whose results are briefly stated here by Helmholtz, will be found in the paper 'On the Facts Underlying Geometry'. For more detailed criticism of Helmholtz' presuppositions – which however is of no importance for the epistemological outcome – we may refer to the comments on that paper.

[30] The characteristic of a fixed body (the unchanging separation between two points in the body) cited by Helmholtz in what follows, gives the basis on which every physical measurement must rest, for in the last analysis a measurement always occurs through repeatedly applying a measuring rod. But a measuring rod is a body on which two (or more) points are marked whose separation is regarded as constant. To 'apply' the measuring rod to an object to be measured means: to bring those two points into coincidence with specific points in the object.

[31] This definition reduces congruence (the equality of two extents*) to the coincidence of point pairs in rigid bodies "with the same point pair fixed in space", and thus presupposes that 'points in space' can be distinguished and held fixed. This presupposition was explicitly made by Helmholtz in the preceding paragraph, but for this he had to presuppose in turn the existence of "certain spatial structures which are regarded as unchangeable and fixed" (p. 15). Unchangeability and fixity (the term 'rigidity' is more usual nowadays) cannot for its own part again be specified with the help of that definition of congruence, for one would otherwise clearly go round in a circle. For this reason the definition seems not to be logically satisfactory.

One escapes the circle only by stipulating by convention that certain bodies are to be regarded as rigid, and one chooses these bodies such that the choice leads to a simplest possible system of describing nature (Poincaré, *Der Wert der Wissenschaft* [*op. cit.*], pp. 44f.) It is easy to find bodies which (if temperature effects and other influences are excluded) fulfil this ideal sufficiently closely in practice. Then congruence can be defined unobjectionably (as by Einstein in *Geometrie und Erfahrung* ['Geometry and Experience'], p. 9) as follows: "We want to term an extent what is embodied by two marks made on a body which is in practice rigid. We think of two practically rigid bodies with an extent marked on each. These two extents shall be called 'equal to each other' if the marks on the one can constantly be brought into coincidence with the marks on the other."

[32] This is the so-called monodromic principle. Compare on this point the second paper in this volume, and in particular the comments upon it.

[33] 'Hamilton's principle' expresses the laws of nature in such a general and comprehensive form that it has also held good for all modern extensions of physical conceptions. According to it, the course of all physical processes is such that a certain function of the magnitudes that determine the momentary state takes a minimum value (in exceptional cases a maximum one) for the transition from the initial to the final state.

---

* ['Strecken', i.e. 'stretches' or 'linear extensions'; the same word is used by Einstein in the quotation below.]

[34] 'Necessities of thought' . The results of our thought should only be called *necessary* if they have been obtained by deduction (compare note 1). But since the particular specifications of our in fact existing space cannot be deduced from the mere concept of a three-dimensional manifold, Helmholtz rightly denies that they are necessary for thought. In Kant's terminology this conclusion would have to be formulated by saying: the propositions which assert those particular properties of our space – i.e. the axioms of geometry – are not analytic. For indeed those judgments were called analytic which only assert of an object what is already contained in its concept, thus what can be deduced from it.

[35] The 'contrary assumption' would, in the first instance, be only that the axioms of geometry constitute *synthetic* propositions, for this is the contrary of analytic, and the latter coincides, according to the preceding note, with 'necessary for thought'. But Helmholtz already goes one step further and raises the question of whether the axioms are empirical propositions, i.e. synthetic *a posteriori* judgments. He therefore seems to have in view only these two possibilities for the axioms: either analytic, or synthetic *a posteriori*. (Thus already above: compare note 12). He seems from the outset not to allow for the very class of judgments on which everything depends here, namely the synthetic *a priori* ones.

Empirical propositions, which Helmholtz opposes to judgments necessary *for thought*, thus to analytic judgments, are in fact the contrary of *necessary* judgments as such; as against this, Kant's synthetic *a priori* judgments, if there were such, would indeed not be necessary *for thought*, but they would constitute *intuitive* necessities. Even according to Kant's doctrine, non-Euclidean axiom systems would be wholly *thinkable*, i.e. they could be set up without contradiction. But they would not be *imaginable*, not realizable in products of intuitive imagination, and they could find no application in physical actuality, for this, as being intuitively perceptible, would be subject to the laws governing the manner in which we intuit. If the Euclidean axioms were amongst these laws, then we would be unable to perceive and imagine the corporeal world other than as ordered in Euclidean space.

Kantians have objected against Helmholtz that he did not distinguish between intuitive necessity and necessity for thought, and indeed he omitted to mention this important distinction here. But he was fully aware, in fact, that the resolution of his problem depends precisely upon whether some other geometry is *intuitively imaginable* besides Euclidean geometry. This already emerges very clearly from the next sentence of the text, for there Helmholtz uses the word 'imagine', whose sense he explicitly defined above (compare note 15). He rightly attaches great importance to that definition, for he comes back to it several times (p. 23 of this lecture, also in "The Facts in Perception", below p. 130; likewise in his *Abhandlungen*, vol. II, p. 644).

[36] This sentence formulates the problem precisely and is completely correct. If "spaces of another kind are imaginable in the sense stated" (compare the previous note) then Euclidean space has no intuitive necessity for us; and since Euclid's axioms, as already shown by Helmholtz, are not necessary *for thought* either, proof will have been given that they are not necessary *at all*, and consequently (compare note 2) not valid *a priori*. They will then indeed have to have originated in experience and be *a posteriori*. Helmholtz quickly applies himself in the text to the demonstration that non-Euclidean spaces in fact are imaginable. The following paragraphs start with preparatory considerations.

[37] By the parallax of a fixed star is understood the angle which the major axis of the

earth's orbit subtends as viewed from the star. The fixed star and the two extremities of the axis of the earth's orbit form together the terminal points of a very elongated triangle; the angle at the tip is the parallax. As the sum of angles of a Euclidean triangle equals two right angles, one obtains the value of the parallax by substracting from two right angles the sum of the two angles at the base of the triangle. For immeasurably distant stars, the two long sides of the triangle become markedly parallel, the sum of the base angles becomes equal to two right angles and the parallax zero. As the sum of angles of a pseudospherical triangle is smaller than two right angles, the sum of the base angles cannot reach this total. Thus if one substracts it from two right angles, the result will always have to be positive even for the most distant stars, as Helmholtz mentions in parentheses. In spherical space, on the other hand, one will even get negative values for the angle, since the sum of the two base angles could exceed two right angles.

[38] It holds only with certain reservations that the axiom of parallels, and consequently the Euclidean structure of space, can be "empirically confirmed" by astronomical observations. Above all, it only holds under the presupposition that light rays are to be regarded as straight lines. If the determinations of parallax e.g. yielded negative values, this result could always be explained just as well by the assumption of curvilinear propagation of light as by the hypothesis of positive curvature of space. This has been pointed out with particular emphasis by H. Poincaré, who says with regard to this case (*Wissenschaft und Hypothese* ['Science and hypothesis'], 2nd ed., 1906, p. 74): "one would thus have the choice between two conclusions: we could renounce Euclidean geometry, or alter the laws of optics and allow that light is not propagated exactly in a straight line." It is unquestionably a fault in Helmholtz' account that he does not sufficiently emphasize the second possibility and sharply contrast the two alternatives with each other. He thereby appeared to offer an easy target for the attacks of the Kantians. We shall return to this later, and then see that he perceived the true state of affairs with complete clarity.

Poincaré is thoroughly wrong in continuing at that point: "It is needless to add that everybody would consider the latter solution to be the more advantageous." The successes of Einstein's general theory of relativity, which sacrifices the validity of the Euclidean axioms, prove the error of Poincaré's assertion, and it may be said with certainty that he would today gladly withdraw it in the face of those successes.

He believed that Euclidean geometry would always have to retain its preferential status in physics because it was the 'simplest'. Yet it is not the simplicity of an individual branch or an auxiliary means of science which is decisive, but it ultimately comes down to the simplicity of the *system* of science, a simplicity which is identical with the *unity* of natural knowledge. This maximal simplicity is nowadays attained more perfectly by dropping Euclidean geometry than by retaining it. The observation that the planets do not travel around the sun precisely in simple Keplerian ellipses, but describe extremely complicated orbits, nonetheless made the world picture simpler, because it enabled one to ascribe to Newton's law of gravitation a more precise validity. Just as Poincaré (*ibid.*, p. 152) recognized that "the simplicity of Kepler's laws is only apparent", he would equally say today: the simplicity of Euclidean geometry in its application to nature is only apparent. Helmholtz, attaching himself to Riemann, advocated the admissibility of this standpoint clearly and resolutely; what he contributed as well to its philosophical justification will shortly be considered.

[39] 'Congruence' is established by observing the coincidence of material points. All physical measurements can be reduced to this same principle, since any reading of any of our

instruments is brought about with the help of coincidences of movable parts with points on a scale, etc. Helmholtz' proposition can therefore be extended to the truth that no occurrences whatsoever can be ascertained physically other than meetings of points, and from this Einstein has logically drawn the conclusion that all physical laws should contain basically only statements about such coincidences. The following paragraphs of Helmholtz' lecture contain statements going in the same direction.

⁴⁰ In the little word 'actually' there lurks the most essential philosophical problem of the whole lecture. What kind of sense is there in saying of a body that it is *actually* rigid? According to Helmholtz' definition of a fixed body (p. 15), this would presuppose that one could speak of the distance between points 'of space' without having regard to bodies; but it is beyond doubt that without such bodies one cannot ascertain and measure the distance in any way. Thus one gets into the difficulties already described in note 31. If the content of the concept 'actually' is to be such that it can be empirically tested and ascertained, then there remains only the expedient already mentioned in that note: to declare those bodies to be 'rigid' which, when used as measuring rods, lead to the *simplest* physics. Those are precisely the bodies which satisfy the condition adduced by Einstein (compare note 31). Thus what has to count as 'actually' rigid is then not determined by a logical necessity of thought or by intuition, but by a convention, a definition.

⁴¹ Here it seems as if Helmholtz did, after all, regard the concept of a rigid measuring rod as something absolute, continuing to exist independent of our conventions, for one can speak of the probability or truth of an assertion only if its correctness can in principle be tested; the concept of probability is not applicable to definitions. But we obviously have to see in this formulation only a concession to the reader's capacity of comprehension, and Helmholtz only gradually introduces him to the rigorous consistency of the more radical thoughts.

⁴² This behaviour is in fact not specified immediately, but only after three purely preparatory paragraphs, in the one which follows them.

⁴³ With regard to what follows, compare the presentation of the same thoughts in Delboeuf, *Prolégomènes Philosophiques de la Géométrie* ['Philosophical prolegomena to geometry'], 1860; Mongré, *Das Chaos in kosmischer Auslese* ['Chaos in cosmic selection'], 1898, ch. 5; Poincaré, *Der Wert der Wissenschaft*, [*op. cit.*], pp. 46f.; *Science et Méthode* ['Science and method'], 1908, pp. 96ff.; *Dernières Pensées* ['Last thoughts'], 1913, pp. 37ff.; Schlick, *Raum und Zeit* ['Space and time'], 3rd ed., 1920, ch. 3.

⁴⁴ The hypothetical change described here by Helmholtz can be expressed in mathematical language thus: if one thinks of the whole world deformed such that its new shape arises from its old one by a one-to-one and continuous, but otherwise quite arbitrary point transformation, then the new world is in actuality completely indistinguishable from the old. The mirror example which follows in the text corresponds to a special case of such a transformation. (Compare the literature cited in note 43.)

⁴⁵ The realization that the question is meaningless as to which of the two worlds described is the 'actual' and which the deformed one, is of the highest importance for the whole problem. There exists no 'actual' difference at all between the two worlds, but only one of description, in that a different kind of coordinate system is assumed in each case (com-

pare again the literature cited above in note 43). In other words: in both cases we are dealing with the same objective actuality, which is portrayed by two different systems of symbols.

[46] If Helmholtz wanted to say, in the last clause of this sentence, that the men of the two worlds would ascertain differing mechanical laws, then this would obviously be an error, so long as he does not drop the presupposition that in the world of the mirror all measuring rods participate in the distortions of the bodies. For otherwise all measurements there, all readings of instruments, must lead to precisely the same numbers as in the original world; all point coincidences remain constant (compare note 39). But a law is only a summary expression for the results of measurements; consequently the physical laws in the two worlds are not different. Yet perhaps Helmholtz is thinking in terms of the observer in the deformed world somehow getting offered measuring rods which do not participate in the deformation otherwise prevailing there, and with which he can continually compare his bodies. In this case he would ascertain peculiar laws of motion, and he could infer that it would be more reasonable to regard the other world as undistorted, because there (with respect to those measuring rods) a much simpler mechanics prevails.

[47] F. Klein has introduced in a purely projective way the specification of the non-Euclidean metric in the interior of a second degree surface (*Ges. Abhandlungen*, vol. I, pp. 287f.).

[48] Wherever the observer may be situated, there always exists a mapping into the Euclidean sphere such that his situation corresponds to the centre of the sphere. Let P be a point in the non-Euclidean space and Q its image in the space of the sphere. If now AA' are the eyes of the observer in the non-Euclidean space and BB' are their images, which are taken to be very close to each other, then the non-Euclidean angles PAA' and $PA'A$ must, apart from infinitely small quantities of higher order, be equal to the angles QBB' and $QB'B$. Thus someone who has Euclidean habits will see Q at the very distance which it in fact has in Beltrami's spherical image.

[49] In this paragraph Helmholtz has formulated the chief epistemological result of his investigation. Having shown, in the preceding paragraphs, that the sense impressions which one would receive in a non-Euclidean world can very well be imagined intuitively, he infers that Euclidean space is *not* an inescapable form of our faculty of intuition, but a product of experience. Is this inference really cogent?

This has been particularly contested on the part of Kantian philosophy (compare for instance A. Riehl, *Helmholtz in seinem Verhältnis zu Kant* [*op. cit.*], pp. 35ff.), with the objection that the various series of perceptions described by Helmholtz need not necessarily be interpreted as processes in a non-Euclidean space, but that one can equally well, as it clearly emerges precisely from Helmholtz' own account, conceive of them as peculiar changes of place and shape in Euclidean space. And this latter kind of interpretation – thus the *a priori* school goes on to assert – is the one which our intuition inevitably forces upon us. It obliges us, if any strange behaviour of bodies is observed, to make their *physical* constitutions and laws alone responsible for this, and forbids us to seek instead the ground for that behaviour in a constitution of 'space' which deviates from the Euclidean one.

We have already emphasized more than once that actually both possibilities of interpretation exist, one beside the other. But Helmholtz was so firmly convinced that we would choose the *geometrical* interpretation for certain experiences of perception and measurement, that he did not explicitly refute here the other possibility. He was himself obviously so capable of freeing himself from rooted habits of intuition, that self-observation seemed to show him immediately the absence of the alleged pure intuition. Yet, probably not everybody can struggle through so easily to this attitude; and thus a gap is to be found here in the reasoning. One should have refuted the *a priori* view according to which we have available, for the interpretation of actuality, only physical hypotheses and not purely geometrical ones, because the latter, though possible in thought, are excluded by an intuitive constraint.

The gap in the proof may be variously filled in. Thus Poincaré makes it extremely plausible, in his books on epistemology, that for instance those lines which we call *straight* are certainly not distinguished over and above other lines by immediately given intuitive properties, but solely by the role which, according to experience, they play in physical actuality; and this result is then in fact decisive. Furthermore, and perhaps above all, it might be pointed out that we do not possess a *purely* geometrical intuition at all, in that for instance surfaces without any thickness, colour, etc. are not imaginable at all. (Compare also F. Klein's lecture of 14 October 1905 to the Philosophical Society of Vienna University; M. Pasch, *Vorlesungen über neuere Geometrie* ['Lectures on modern geometry']. Also already J. S. Mill, and before him Hume in the *Treatise*, Book I, part II, section IV.) Then only physical-corporeal structures are accessible at all to our sensible-intuitive imagination, and it follows that the physical and the geometrical are already inseparably fused together in our intuitive geometry, so that it therefore coincides with 'practical geometry'. This was undoubtedly Helmholtz' opinion. Thus even the gap found here in the proof is, in this sense, closed by him in Appendix III to the address on 'The Facts in Perception'; see below, p. 153. Compare on the question of the existence of a pure intuition also Schlick, *Allgemeine Erkenntnislehre* ['General Epistemology'], 1918, § 37.

[50] On this point a number of mathematicians go beyond Helmholtz, and are of the opinion that we cannot be denied off-hand the capacity to imagine a world of four spatial dimensions. Thus, for example, Poincaré. He points out (*Wissenschaft und Hypothese* [*op. cit.*], p. 70) that the three-dimensionality of our visual space arises, *inter alia*, from the fact that the simultaneous perceptions of one eye, which are basically only two-dimensional, follow each other according to quite specific laws – namely the laws of perspective. Following this he thinks that we would automatically imagine the world to have four spatial dimensions, if observation revealed a certain transformation of three-dimensional corporeal forms, taking place according to quite specific laws. These forms would then have to appear to us as three-dimensional perspective views of four-dimensional structures, which would therefore be intuitively imaginable in the same sense as the third dimension of visual space is for a one-eyed person.

If this is already true for visual space, to which Helmholtz of course restricts his considerations, then for the other ideas of the senses – such as those of the sense of touch and the muscular sense – it seems much less certain still that they must necessarily appear arranged in a three-dimensional manifold. At any rate, the question of the imaginability of a fourth spatial dimension must at least be regarded as a problem deserving serious consideration.

[51] This sentence contains the fundamental insight – which Riemann had already attained and which has recently become truly fruitful in the general theory of relativity – that geometry is to be regarded as a part of physics, and thus not as a science of purely ideal structures, such as say arithmetic, the theory of numbers. The sentences of the text which follow give an irreproachable foundation for this insight.

The sentence is perhaps not quite happily formulated, inasmuch as Helmholtz seems in it to contrast geometry with a 'pure theory of space', yet without stating how the concept of a pure theory of space is to be delimited. Today we would rather say that precisely the 'theory of space' is simply an empirical science, and namely, to use Einstein's way of putting it, that part of physics which deals with the 'situabilities' * of bodies. Even Newton already thought the same, as is proved by the quotation which Helmholtz gives elsewhere (p. 163 of this volume). The purely mathematical discipline of geometry, as opposed to practical or physical geometry, is nothing but a structure of theorems, which are derived in a purely logico-formal way from a series of axioms, without regard to whether there exist any objects (e.g. spatial structures) for which those axioms hold (compare Schlick, *Allgemeine Erkenntnislehre* [*op. cit.*], §7). This rigorous 'geometry' therefore, properly speaking, bears its name without justification, and can raise no claim to be a 'theory of space'. That there indeed exists no 'pure' theory of space was also Helmholtz' opinion, as can easily be seen from his remarks on pp. 155 ff.

52 Compare note 46.

53 In contrast to less careful statements (compare notes 31, 40, 41), this paragraph shows how clear Helmholtz was about the interdependence between geometry and physics: he rightly declares that we can conceive the in fact existing space of our world to be an arbitrary non-Euclidean one, provided only we introduce, at the same time, entirely new laws of a physical kind. Why don't we do this? Why do we in practice, e.g. in technology, always employ Euclidean geometry? Kant would have answered: because our *a priori* spatial intuition does not let it be otherwise! Poincaré would say: because geometry thereby becomes the science of the situabilities of our fixed bodies, and our physical formulae thereupon assume their simplest form. This is naturally also Helmholtz's opinion; for him too, as the first line of this paragraph shows, what decides the conception to be chosen is scientific *usefulness*.

54 This sentence says it once again: the theory of space is not a purely mathematical discipline, but a theory of the situabilities of rigid bodies, or, as Einstein also terms it, 'practical geometry'. Helmholtz also calls it *physical* geometry (see below, p. 154).

55 Helmholtz does not use the word 'transcendental' correctly in its Kantian sense; in particular he often confuses it, as also in this passage, with '*a priori*'.
The sense of the sentences which follow should be further elucidated with a few words in view of the importance of their content. Having rejected the existence of an *a priori* intuition of 'space', Helmholtz examines the question of whether we perhaps possess an *a priori* intuition of the 'fixed body'. What, he asks, would be the consequence? The propositions of the then existing intuitive geometry, though indeed *a priori*, would be *analytic*, for they would be deducible from the properties of those ideal rigid bodies with whose situabilities this geometry would deal. But as to what are the situabilities of any *actual* bodies – on this question of practical geometry it could indeed teach us nothing; the propositions of the latter would continue to be synthetic *a posteriori* judgments.
Let us assume for a moment that this supposed *a priori* geometry were Euclidean, then the only consequence would be that we would be allowed to *call* rigid only those bodies

---

* ["Lagerungsmöglichkeiten"].

whose situational laws obey Euclidean rules. However, it is a purely empirical fact if our most convenient mesuring rods actually are rigid in this sense. This *a priori* intuition would therefore be superfluous, it would contribute nothing to the knowledge of actuality and thus would not fulfil the very task assigned to it by Kant. It would indeed even obstruct and be harmful for our knowledge, as soon as it turned out (as is evidently the case now-adays) that we can only maintain the assumption of strictly Euclidean situabilities if we sacrifice for this the beauty and perfection of our system of physics. In short, the assump-tion of an *a priori* intuition of the rigid body would be scientifically mistaken in every respect. Helmholtz also follows similar trains of thought below on pp. 155 ff.

[56] This paragraph repeats once again that the geometry applied to actuality – which arises from the formally abstract discipline as soon as one brings it into coincidence with actuality at any point whatsoever, and gives its empty concepts a real content – that this practical geometry is an *empirical science* throughout.

CHAPTER II

ON THE FACTS UNDERLYING GEOMETRY[1]

From the *Nachrichten von der königlichen Gesellschaft
der Wissenschaften zu Göttingen*, no. 9, 3 June 1868.
Reprinted in *Wissenschaftliche Abhandlungen*, vol. II, pp. 618–639.

My investigations on spatial intuitions in the field of vision induced me also to start investigations on the question of the origin and essential nature of our general intuitions of space. The question which then forced itself upon me, and one which also obviously belongs to the domain of the exact sciences, was at first only the following: how much of the propositions of geometry has an objectively valid sense? And how much is on the contrary only definition or the consequence of definitions, or depends on the form of description? In my opinion, this question is not to be answered all that simply. For in geometry we deal constantly with ideal structures, whose corporeal portrayal in the actual world is always only an approximation to what the concept demands, and we only decide whether a body is fixed[†], its sides flat and its edges straight, by means of the very propositions whose factual correctness the examination is supposed to show[2].

In this investigation, I had essentially started along the same path as that followed by Riemann in his recently published[*] inaugural dissertation[4]. The *analytic* treatment of the question, of what distinguishes space from other measurable, severally extended and continuous magnitudes, recommends itself in this case precisely because of the circumstance that it lacks an intuitive character. For this reason it is not exposed to the deceptions which, on account of the particular limitations of our intuitions, are so difficult to avoid in this field. Furthermore it has the advantage, that it easily allows one to get a complete view of the possibility of deductively implementing a deviating system of axioms.

So this was my immediate purpose, as Riemann's: to investigate which properties of space belong to any manifold which depends on several variables and goes over continuously into itself, and whose differences are all quantitatively comparable amongst themselves; and which properties, on the other hand, are not conditioned by this general character,

---

[†] ['fest'; see note I.31.]
[*] *Abhandlungen der königlichen Gesellschaft der Wissenschaften zu Göttingen*, vol. XIII[3].

but are peculiar to space. Precisely in physical optics, two examples were available to me of other manifolds which can be portrayed spatially and are variable in several respects. Namely the colour system, which Riemann[5] also cites, and the measuring out of the visual field by visual estimation. Both show certain fundamental differences from the metrical system of geometry, and stimulated comparison.

The priority of a series of the results of my own work has been anticipated by the publication of Riemann's investigations. Even so, I must incidentally confess that it was no small matter to me to see, concerning a subject so unusual and more discredited than not by earlier attempts, that such an outstanding mathematician had honoured the same questions with his interest, and that it was a significant guarantee of the correctness of my chosen path to meet him upon it as a companion. But our separate work does not coincide completely. So I will allow myself here to present to the Königliche Gesellschaft that part of my investigations which is not included in those of Riemann.

Riemann explains that a manifold is to be regarded as n-fold extended when in it the specified individual (the place) can be specified through the specification of n variable magnitudes (coordinates), and he adds the further demand that every line, independent of place and orientation, should be comparable[†] in length with every other[6]. Having done that, he is faced with the problem of determining how the element of length of a line is dependent upon the corresponding differentials of the coordinates. This he does by means of a hypothesis, setting the element of length of the line equal to the square root of a homogeneous second degree function of the differentials of the coordinates. His foundation for this hypothesis is its being the simplest one corresponding to the conditions of the problem. But he explicitly acknowledges it to be a hypothesis, and in particular mentions the possibility that one might also posit for the line element a fourth root of a homogeneous fourth degree expression, or other even more complicated expressions.

Then he deals further, in the most general form, with the consequences to be drawn from that hypothesis. Only at the end does he again specialise this generality, laying down the further demand that bounded n-fold extended structures of finite magnitude (fixed point systems) should be able

† ['vergleichbar'; in the third paper 'vergleichen' is translated by 'liken' instead of 'compare': see Translator's Note.]

to move about everywhere without stretching. This then leads him to the case of actual space, which satisfies this demand. It here emerges, however, that the postulates taken as his basis do not comprehend the demand, laid down in customary geometry, that the dimensions of space be infinite.

My own investigation differs from Riemann's in the following respect. I looked more closely at this finally introduced restriction, which distinguishes actual space from other severally extended manifolds, and namely at its influence on giving a foundation for the proposition upon which the whole investigation hinges, according to which the square of the line element is a homogeneous second degree function of the differentials of the coordinates. It can be shown, that if one rather adheres from the outset to the demand that figures fixed in themselves should have unconditionally free mobility, without distortion, in all parts of space, then Riemann's initial hypothesis is derivable as a consequence of much less restricted assumptions.

My starting point was that the primary measurement of space is entirely based upon the observation of congruence; clearly, the rectilinearity of light rays is a physical fact supported by particular experience in another field, and has no importance whatever for someone who is blind, though he can also attain a complete conviction of the correctness of geometrical propositions. However, one cannot at all speak of congruence unless fixed bodies or point systems can be moved up to one another without changing form, and unless the congruence of two spatial magnitudes is a fact whose existence is independent of all motions. So I presupposed from the outset that the measurement of space through ascertaining congruence was possible, and set myself the task of looking for the most general analytical form of a severally extended manifold in which motions of the kind thus demanded are possible.

With this change of approach, my work lacks the great generality attained by Riemann's analysis before the introduction of the restriction mentioned above. After its introduction my results agree with his completely.

1. The hypotheses underlying the investigation

I. Space of n dimensions is an n-fold extended manifold. In other words, the individual specified in it, the point, is specifiable by measuring any continuously and independently varying magnitudes (coordinates)[7], whose total number is n. Thus any motion of a point is accompanied by a continuous change of at least one of the coordinates. Should there be exceptions, where either the change becomes discontinuous or despite the motion no change whatever takes place in any of the coordinates, then these exceptions are at least restricted to certain locations limited by one or several equations (thus to points, lines, surfaces and so on), and these locations can initially remain excluded from the investigation.

It is to be noted, that by continuity of change during motion we do not merely mean that all values of the changing magnitudes intermediate between the terminal values are run through, but also that derivatives exist. In other words, the ratios of associated changes of the coordinates approach a fixed ratio, as the magnitude of these changes progressively decreases.

This hypothesis also underlies Riemann's work, and I may refer to it for a more detailed discussion and foundation.

II. The existence of mobile but rigid bodies, or point systems, is presupposed, such as is needed to enable the comparison of spatial magnitudes by congruence. As we are not yet allowed to presuppose here any special methods of measuring spatial magnitudes, the definition of a fixed body can only be the following at this point: between the $2n$ coordinates of any point pair belonging to a body fixed in itself, there exists an *equation* which is *independent* of the motion of the latter, and which for all congruent point pairs is *the same*. While point pairs are congruent, if they can coincide simultaneously or successively with the same point pair of the space.

Although its formulation seems so inspecific, this definition of a fixed body has extremely far-reaching implications, because according to it there must exist $m(m-1)/2$ equations between m points, while the number of unknowns contained in them, i.e. the number of coordinates, is mn. Moreover a number of the latter, namely $n(n+1)/2$, must remain at our disposal, corresponding to the variable situation of the fixed system. So when $m > n+1$, we have here $\frac{1}{2}(m-n) \times (m-n-1)$ more equations than unknowns. It follows from this that not just any arbitrary kind of equations can exist between the coordinates of each pair of fixed points, but that these equations instead have quite particular properties. There thus arises from this the specific analytical problem of specifying more precisely what kind of equations these are.

I will note that the postulate laid down above, according to which in space there exists an equation between *every pair* of fixedly connected points, distinguishes space from the colour system. In the latter there exists in general, on account of the law of mixtures, an equation only between five points and not less. Or in the special case in which a colour can be mixed together from two others, between these three. The case corresponding to this in space would be if all fixed bodies could be expanded arbitrarily in the directions of three principal axes. The definition of fixity given above is therefore the definition of the highest conceivable degree of relative fixity[8].

III. *Completely free* mobility of fixed bodies is presupposed. It is presupposed, in other words, that any point in them can move continuously to the place of any other, to the extent that it is not restricted by the equations that exist between it and the remaining points of the fixed system to which it belongs.

Thus given a system fixed in itself, its first point is absolutely mobile. When this point is determined, there exists an equation for the second point, and one of its coordinates becomes a function of the remaining $(n-1)$. After the second point too has been determined, there exist two equations for the third, and so on. Thus one requires altogether $n(n+1)/2$ magnitudes in order to specify the situation of a system fixed in itself[9].

The following is a consequence of this assumption and the one laid down under II. If two point systems A and B, fixed in themselves, could be brought into a congruence of corresponding points for an initial situation of A, then given any other situation of A, they must still be capable of being brought into a congruence of exactly the same points as were congruent before. That means, in other words, that the congruence of two spatial structures[10] is not dependent upon their situation. Or that all parts of space, if one disregards their limits, are mutually congruent, just as all parts of the same spherical surface, disregarding their limits, are congruent with each other in respect of the curvature of the surface[11].

The field of vision reveals a more restricted mobility of retinal images on the retina [12]. As to what special consequences derive from this for the measurement of distances by visual estimation, these I have discussed in my physiological optics [13].

IV. We must finally attribute to space one more property, which is analogous to the monodromic property of functions of a complex magnitude. It expresses itself in the circumstance, that two congruent bodies are also still congruent after one of them has undergone a complete rotation about any axis of rotation. The analytic characterisation of *rotation*, is that a certain number of points in the moving body keep their coordinates unchanged during the motion. That of *reversal* of the motion, is that continuously transforming value complexes of the coordinates which were run through earlier, are run through backwards. We can express the fact concerned as follows: If a fixed body rotates about $(n-1)$ of its points, and these are so chosen that its position then depends only upon one independent variable, then rotation without reversal finally *returns it to the initial situation* from which it started [14].

We shall see that this last property of space need not necessarily be present, even if our first three conditions are fulfilled. So however self-evident it may seem, it had to be adduced as a particular property. ·

Geometry as we know it silently presupposes this last property, when it treats the circle as a closed line. It presupposes postulates II and III for the congruence theorems, since the existence of bodies fixed in themselves but freely mobile, having the properties stated there, is the precondition of any congruence. It equally presupposes the continuity and dimensions of space. These propositions have been adduced here in analytical form alone, as without employing such a form their sense simply cannot be precisely expressed.

2

The consequences of these initial propositions will be drawn by me for the case of three dimensions. I also note, that as it will only be a matter in what follows of giving a foundation for Riemann's proposition concerning the differentials of the coordinates I shall use the assumptions II, III and IV only for points having infinitely small coordinate differences. So this congruence independent of limits will be presupposed only for infinitely small spatial elements[15].

Let u, v, w be the coordinates of a point belonging to a fixed body, in the first situation of this body[16]. Let r, s, t be the coordinates of the same point in a second situation of the fixed body. These will have to be functions of u, v, w and six constants (position constants) which specify the new situation of the fixed body. Then r, s and t, according to assumption I, will have to vary continuously with u, v, w, excepting any places, if there be such, where a motion of the point produces discontinuous variations of the coordinates. So where this is not the case, we shall have:

$$\left. \begin{aligned} du &= \frac{du}{dr}\,dr + \frac{du}{ds}\,ds + \frac{du}{dt}\,dt \\[2mm] dv &= \frac{dv}{dr}\,dr = \frac{dv}{ds}\,ds + \frac{dv}{dt}\,dt \\[2mm] dw &= \frac{dw}{dr}\,dr + \frac{dw}{ds}\,ds + \frac{dw}{dt}\,dt \end{aligned} \right\} . \tag{1}$$

In these the derivatives are functions of u, v, w, or of r, s, t which are dependent upon them, and besides this are functions of the six position constants.

The functional determinant of u, v, w will not be able to vanish here, with the possible exception of places where either u, v, w or r, s, t are not sufficient to specify the complete situation of a point.

On the other hand, let the fixed body go over from the first situation, where the coordinates of its points were u, v, w, to a third where they are ρ, σ, τ. We shall then as before have:

$$\left. \begin{aligned} du &= \frac{du}{d\rho}\,d\rho + \frac{du}{d\sigma}\,d\sigma + \frac{du}{d\tau}\,d\tau \\[2mm] dv &= \frac{dv}{d\rho}\,d\rho + \frac{dv}{d\sigma}\,d\sigma + \frac{dv}{d\tau}\,d\tau \\[2mm] dw &= \frac{dw}{d\rho}\,d\rho + \frac{dw}{d\sigma}\,d\sigma + \frac{dw}{d\tau}\,d\tau \end{aligned} \right\} . \tag{1a}$$

Here too the functional determinant will not be able to be zero, in both cases excluding the same exceptions as above.

Now of the six constants which specify the position of the fixed body in the second situation, we shall be able to choose three such that the situation of the point u, v, w is the same in the second position of the system, as that of the same point in the third position[17] (assumption III). So that then:

$$r = \rho, \qquad s = \sigma, \qquad t = \tau.$$

If one now takes the values of du, dv, dw from Equation (1) and puts them in Equation (1a), one then obtains linear and homogeneous expressions for dr, ds and dt in $d\rho$, $d\sigma$ and $d\tau$, or for the latter in the former. Since, as noted, the determinants of Equations (1) and (1a) cannot become zero, provided the coordinates are sufficient to specify the situation of the points concerned, then one can always, given this presupposition, set up such a linear expression. We can write it:

$$\left. \begin{aligned} dr &= A_0 d\rho + B_0 d\sigma + C_0 d\tau \\ ds &= A_1 d\rho + B_1 d\sigma + C_1 d\tau \\ dt &= A_2 d\rho + B_2 d\sigma + C_2 d\tau \end{aligned} \right\} \cdot \tag{2}$$

It must be possible, disregarding the special exceptional cases mentioned, to set up linear expressions of this kind. This results from the fact, that the point r, s, t under consideration has no specially distinguished relation, supposed in the nature of the problem to u, v, w, but is quite arbitrary. Equally the point ρ, σ, τ. Thus equations (1) and (1a) must be correct in the general case, and from them there follows (2). One would not put down (2) directly with the same assurance. For in a motion during which the point r, s, t stayed put, this would indeed stand in a preferred relation to ρ, σ, τ, possibly making it doubtful whether the first derivatives might not collectively vanish.

The point which in the first situation has the coordinates $u + du$, $v + dv$, $w + dw$, has in the second situation the coordinates $r + dr$, $s + ds$, $t + dt$ and in the third the coordinates $\rho + d\rho$, $\sigma + d\sigma$, $\tau + d\tau$. Thus the magnitudes dr, ds, dt refer to the same point, in another situation of the system to which it belongs, as $d\rho$, $d\sigma$, $d\tau$. Equations (2) express the most general law which must exist between these magnitudes, if three-dimensional space is to be measurable by means of three continuously varying magnitudes.

In what follows I shall introduce the notation [18]:

$$\begin{aligned}
dr &= \varepsilon x & d\rho &= \varepsilon \xi \\
ds &= \varepsilon y & d\sigma &= \varepsilon \upsilon \\
dt &= \varepsilon z & d\tau &= \varepsilon \zeta
\end{aligned} \left.\vphantom{\begin{aligned} dr \\ ds \\ dt \end{aligned}}\right\} \cdot \tag{2a}$$

Here ε is to refer to a vanishingly small magnitude. We then have:

$$\begin{aligned}
x &= A_0 \xi + B_0 \upsilon + C_0 \zeta \\
y &= A_1 \xi + B_1 \upsilon + C_1 \zeta \\
z &= A_2 \xi + B_2 \upsilon + C_2 \zeta
\end{aligned} \left.\vphantom{\begin{aligned} x \\ y \\ z \end{aligned}}\right\} \cdot \tag{2b}$$

In these equations, the coefficients A, B, C depend upon the three [19] position constants still at our disposal, which specify the position of the system in the last situation. We will denote them by p', p'' and p'''. If we let these constants vary by the vanishingly small magnitudes dp', dp'', dp''', then the second situation of the system varies, and with it the values x, y, z by dx, dy, dz. If we now denote a new variable by η, and put for the presupposed small displacement:

$$\mathfrak{A}_n d\eta = \frac{dA_n}{dp'} dp' + \frac{dA_n}{dp''} dp'' + \frac{dA_n}{dp'''} dp''' \tag{3}$$

and give the letters $\mathfrak{B}_n$ and $\mathfrak{C}_n$ the meaning which corresponds for the B and C, there results [20]:

$$\begin{aligned}
\frac{dx}{d\eta} &= \mathfrak{A}_0 \xi + \mathfrak{B}_0 \upsilon + \mathfrak{C}_0 \zeta \\
\frac{dy}{d\eta} &= \mathfrak{A}_1 \xi + \mathfrak{B}_1 \upsilon + \mathfrak{C}_1 \zeta \\
\frac{dz}{d\eta} &= \mathfrak{A}_2 \xi + \mathfrak{B}_2 \upsilon + \mathfrak{C}_2 \zeta
\end{aligned} \left.\vphantom{\begin{aligned} \frac{dx}{d\eta} \\ \frac{dy}{d\eta} \\ \frac{dz}{d\eta} \end{aligned}}\right\} \cdot \tag{3a}$$

If in these equations we express ξ, υ, ζ from (1) and (1a) and (2a) linearly in terms of x, y, z, which must always be possible according to what was said above, we then obtain expressions of the form [21]:

$$\left.\begin{array}{l} \dfrac{dx}{d\eta}=a_0x+b_0y+c_0z \\[2mm] \dfrac{dy}{d\eta}=a_1x+b_1y+c_1z \\[2mm] \dfrac{dz}{d\eta}=a_2x+b_2y+c_2z \end{array}\right\} . \tag{3b}$$

Since each of the magnitudes a, b, c includes three of the arbitrarily variable magnitudes dp', dp'', dp''', there can be an infinite number of such transformation systems. But given the coefficients of any four of them, it will always be possible, by eliminating dp', dp'' and dp''', to obtain between these coefficients a system of linear equations:

$$a_n=fa'_n+ga''_n+ha'''_n$$
$$b_p=fb'_p+gb''_p+hb'''_p$$
$$c_q=fc'_q+gc''_q+hc'''_q$$

where f, g, h are constants and n, p, q refer to any of the indices 0, 1, 2.

If the systems a'_0 etc., a''_0 etc., a'''_0 etc. are themselves such that, between their coefficients, there exists no system of equations such as that given above, then *every other* system corresponding to a possible motion will be expressible linearly in terms of the coefficients a', a'', a''' etc., and every sum having the form of the expressions above for a_n, b_p and c_q with arbitrary constants f, g, h will correspond to a possible motion.

There exists another specification for the various motions of this kind. For according to assumption III, after one point of the system r, s, t has been determined, one can still determine any other point as being at rest, without thereby making further motion impossible. We must therefore be able to vary the magnitudes dp', dp'' and dp''' such that for arbitrarily given values of x_0, y_0, z_0 there can arise [22]:

$$0=a_0x_0+b_0y_0+c_0z_0$$
$$0=a_1x_0+b_1y_0+c_1z_0$$
$$0=a_2x_0+b_2y_0+c_2z_0 .$$

This can only occur if, for all infinitely small rotations of the system, the following condition is satisfied for the determinant of the coefficients:

$$\begin{vmatrix} a_0 & b_0 & c_0 \\ a_1 & b_1 & c_1 \\ a_2 & b_2 & c_2 \end{vmatrix} = 0. \tag{4}$$

The first infinitely small displacement, whereby η has gone over into $\eta + \delta\eta$, x into $x + dx$, y into $y + dy$ and z into $z + dz$, can be followed by a second one of the same kind and the same magnitude. Let us call the system in its first situation A_1 and in the second A_2. If we think of both simultaneously existing, then the points $(x + dx, y + dy, z + dz)$ in A_1 coincide with the points in A_2 which originally had the situation (x, y, z). If we now let A_1 undergo the same displacement as that whereby it was originally transformed into A_2, then A_2 will also enter a new situation A_3, and we shall be able to regard η as having increased to $\eta + 2\delta\eta$, and to take the coefficients a, b, c as independent of η. At the same time, according to the last part of assumption III, the points which had the coordinates (x, y, z) before the first displacement, will now reach the same situation as that reached through the first displacement by the points with the coordinates $(x + dx, y + dy, z + dz)$. This we can repeat as often as we wish.

In each such displacement the points with the coordinates (x, y, z), in as precise accordance with the Equations (3b) as the first time, will go over into $(x + dx, y + dy, z + dz)$. So if the same displacements carry on continuously, the coefficients a, b, c of Equations (3b) remain constant, while η increases proportional to time. Moreover if one takes x, y, z in reference to a specified point in the moving system, they will vary in such a way as is prescribed by equations (3b) when one regards $dx/d\eta$, $dy/d\eta$ and $dz/d\eta$ in these as derivatives.

In order to carry out the integration of Equations (3b), we look for four new constants by means of the following equations [23]:

$$\left. \begin{aligned} lh &= la_0 + ma_1 + na_2 \\ mh &= lb_0 + mb_1 + nb_2 \\ nh &= lc_0 + mc_1 + nc_2 \end{aligned} \right\}. \tag{4a}$$

By elimination of l, m, n these give the determinant:

$$\begin{vmatrix} a_0-h & a_1 & a_2 \\ b_0 & b_1-h & b_2 \\ c_0 & c_1 & c_2-h \end{vmatrix}=0 . \qquad (4b)$$

This is a third degree equation in h, which therefore yields three roots. Each of these, when inserted in Equations (4a), gives a system of values for l, m, n in which each time one of these constants remains arbitrary.

If the Equations (4a) are satisfied, then it follows from (3b) that:

$$\frac{d}{d\eta}\{lx+my+nz\}=h\{lx+my+nz\} \qquad (4c)$$

or if we denote the constant of integration by A:

$$lx+my+nz=Ae^{h\eta}. \qquad (5)$$

Moreover, the Equations (4c) and (5) hold for each of the three systems of values yielded by Equations (4a) and (4b).

On account of Equation (4), one of the values of h must be equal to zero. For this one[24]:

$$l_0x+m_0y+n_0z=\text{const.} \qquad (5a)$$

The other two, h_1 and h_2, could be real or complex conjugate magnitudes. In the first case the corresponding l, m, n are also real, in the second case complex.

If the two roots h_1 and h_2 are real, it then follows from the equations of the form (5) that the corresponding magnitudes $(l_1x+m_1y+n_1z)$ and $(l_2x+m_2y+n_2z)$ can vary continously from the value 0 to $\pm\infty$. But they cannot return to an earlier value, as postulate IV demands, without reversal or discontinuity. Therefore the magnitudes x, y, z themselves also cannot do this. The same holds also for the case where h_1 and h_2 are equal in magnitude. One then obtains one linear function of x, y, z which is equal to $e^{h\eta}$, and another one equal to $\eta e^{h\eta}$. The same equally holds when h_1 and h_2 simultaneously become vanishingly small, and so approach very close to the value $h_0=0$. One can then put together three linear functions, of which one is constant, one equal to η and one equal to η^2.

If, on the other hand, h_1 and h_2 have complex values, then that is also the case with the corresponding l, m, n. So if we put:

$$h_1 = \theta + \omega i \qquad\qquad h_2 = \theta - \omega i$$
$$l_1 = \lambda_0 + \lambda_1 i \qquad\qquad l_2 = \lambda_0 - \lambda_1 i$$
$$m_1 = \mu_0 + \mu_1 i \qquad\qquad m_2 = \mu_0 - \mu_1 i$$
$$n_1 = v_0 + v_1 i \qquad\qquad n_2 = v_0 - v_1 i$$

then

$$\lambda_0 x + \mu_0 y + v_0 z = A e^{\theta \eta} \cos(\omega \eta + c)$$
$$\lambda_1 x + \mu_1 y + v_1 z = A e^{\theta \eta} \sin(\omega \eta + c).$$

In this case:

$$(\lambda_0 x + \mu_0 y + v_0 z)^2 + (\lambda_1 x + \mu_1 y + v_1 z)^2 = A^2 e^{2\theta \eta}. \tag{5b}$$

This equation makes it equally impossible for x, y, z to return to their earlier values without reversal or discontinuity, unless $\theta = 0$.

Thus the postulate laid down under IV can only be satisfied, if those roots of Equation (4b) which are not zero are purely imaginary. According to Equation (4b), that occurs when:

$$a_0 + b_1 + c_2 = 0. \tag{6}$$

To determine x, y, z as a function of η, we therefore finally have the three equations:

$$\left. \begin{array}{l} l_0 x + m_0 y + n_0 z = \text{const.} \\ \lambda_0 x + \mu_0 y + v_0 z = A \cos(\omega \eta + c) \\ \lambda_1 x + \mu_1 y + v_1 z = A \sin(\omega \eta + c) \end{array} \right\}. \tag{6a}$$

The determinant:

$$\begin{vmatrix} l_0 & m_0 & n_0 \\ \lambda_0 & \mu_0 & v_0 \\ \lambda_1 & \mu_1 & v_1 \end{vmatrix}$$

cannot become zero without there being an equation which takes η to be constant, and therefore eliminates the motion. Consequently, the magnitudes x, y, z can be determined unambiguously from the three equations (6a), as functions of η.

From here on it will simplify the calculation, if instead of the magnitudes x, y, z we introduce into the calculation the three magnitudes found above:

$$X = l_0 x + m_0 y + n_0 z$$
$$Y = \lambda_0 x + \mu_0 y + v_0 z \quad \biggr\} \; . \tag{6b}$$
$$Z = \lambda_1 x + \mu_1 y + v_1 z$$

From these we can always regain x, y, z as unambiguously determined magnitudes.

So far we have only investigated one kind of rotation, in which a point x_0, y_0, z_0 was to remain fixed. Now according to (6a), in the motion investigated [25]:

$$\frac{dX}{d\eta} = 0$$
$$\frac{dY}{d\eta} = -\omega Z \quad \biggr\} \; . \tag{6c}$$
$$\frac{dZ}{d\eta} = \omega Y$$

The last two magnitudes are therefore equal to zero for those points for which:
$$Y = Z = 0.$$

These are the points which remain at rest in the motion so far considered.

3

We now still have to investigate the other kinds of rotation of the system. As noted above, we can take any other point of the system to be at rest during the rotation.

Let us assume a second rotation, in which the points $X = Z = 0$ stay at rest [26]. If we call the variable which then increases proportional to time η', we can write:

$$\frac{dX}{d\eta'} = \alpha_0 X + 0 + \gamma_0 Z$$
$$\frac{dY}{d\eta'} = \alpha_1 X + 0 + \gamma_1 Z \quad \biggr\} \; . \tag{7}$$
$$\frac{dZ}{d\eta'} = \alpha_2 X + 0 + \gamma_2 Z$$

The middle column of coefficients had to be set equal to zero, because for $X=Z=0$ the derivatives on the left hand side are to become equal to zero.

The two conditions given by Equations (4) and (6), to which every system of coefficients must be subject if there are to be rotations which return into themselves, reduce here to:

$$\alpha_0 + \gamma_2 = 0. \tag{7a}$$

For a third rotation, we take the condition that those points stay in their place for which $X = Y = 0$. Let the variable which increases proportional to time be η'', then we can write:

$$\left. \begin{aligned} \frac{dX}{d\eta''} &= a_0 X + b_0 Y + 0 \\ \frac{dY}{d\eta''} &= a_1 X + b_1 Y + 0 \\ \frac{dZ}{d\eta''} &= a_2 X + b_2 Y + 0 \end{aligned} \right\}. \tag{8}$$

with the condition:

$$a_0 + b_1 = 0. \tag{8a}$$

Now, from the form of the coefficients given in Equation (3), as was already noted there, it results that if two systems of coefficients satisfy the conditions of the problem, then the sum of corresponding coefficients must also give a system satisfying these conditions. If we apply this to (6c) and (7), it follows that:

$$\begin{vmatrix} \alpha_0 & 0 & \gamma_0 \\ \alpha_1 & 0 & \gamma_1 - \omega \\ \alpha_2 & \omega & -\alpha_0 \end{vmatrix} = 0$$

or:

$$\alpha_0 \omega^2 - \omega(\alpha_0 \gamma_1 - \alpha_1 \gamma_0) = 0.$$

As the coefficients of each of these systems contain an arbitrary constant as a factor, we must have separately: $\alpha_0 = 0$, and so $\gamma_2 = 0$ as well, and also $\alpha_1 \gamma_0 = 0$.

Now γ_0 cannot be set equal to zero without clashing with postulate IV, as it would then follow from Equations (7) that:

$$\frac{dX}{d\eta'}=0, \quad \text{therefore} \quad X=C$$

and

$$Z=\alpha_2 C\eta'+C'$$
$$Y=\alpha_1 C\eta'+\gamma_1 C'\eta'+\tfrac{1}{2}\alpha_2\gamma_1 C\eta'^2+C'',$$

where C, C' and C'' are constants. That would represent a rotation which did not return into itself. I will note immediately here that α_2, for the same reason, also cannot become zero.

Since now γ_0 must not be zero, the equation $\alpha_1\gamma_0=0$ allows only the one solution:

$$\alpha_1=0$$

and so the system of coefficients of Equations (7) reduces to:

$$\begin{matrix} 0 & 0 & \gamma_0 \\ 0 & 0 & \gamma_1 \\ \alpha_2 & 0 & 0. \end{matrix}$$

Using the same procedure, it results for the system of coefficients of Equations (8) that:

$$a_0=b_1=0$$
$$a_2 b_0=0.$$

Here a_1 and b_0 must not be zero, for the same reasons as α_2 and γ_0. We must consequently have $a_2=0$, and the system reduces to:

$$\begin{matrix} 0 & b_0 & 0 \\ a_1 & 0 & 0 \\ 0 & b_2 & 0. \end{matrix}$$

Lastly, if one forms the sum of all three systems, one obtains the condition:

$$\begin{vmatrix} 0 & b_0 & \gamma_0 \\ a_1 & 0 & \gamma_1-\omega \\ \alpha_2 & b_2+\omega & 0 \end{vmatrix}=0$$

or:
$$b_0(\gamma_1 - \omega)\alpha_2 + \gamma_0 a_1(b_2 + \omega) = 0. \tag{9}$$

Since, as was noted, this equation must hold even if one multiplies the coefficients belonging to the same system by an arbitrary constant, we must have separately:

$$\gamma_0 a_1 - b_0 \alpha_2 = 0 \tag{9a}$$

$$\left.\begin{array}{l} b_0 \gamma_1 \alpha_2 = 0 \\ a_1 b_2 \gamma_0 = 0 \end{array}\right\}. \tag{9b}$$

Since moreover, as noted, neither b_0 and a_1 nor α_2 and γ_0 can be made equal to zero, we must have:

$$\gamma_1 = 0 \quad \text{and} \quad b_2 = 0.$$

If we put:

$$\alpha_2 = -\phi, \qquad \gamma_0 = \kappa\phi, \qquad a_1 = \psi,$$

it follows from Equation (9a) that:

$$b_0 = -\kappa\psi.$$

From this we now finally obtain the complete system of possible transformations for vanishingly small displacements:

$$\left.\begin{array}{l} dX = -\kappa\psi\, Yd\eta'' + \kappa\phi Zd\eta' \\ dY = \psi Xd\eta'' - \omega Zd\eta \\ dZ = -\phi Xd\eta' + \omega Yd\eta \end{array}\right\}. \tag{10}$$

This contains three arbitrarily variable magnitudes $d\eta$, $d\eta'$ and $d\eta''$, and so must comprehend all possible rotations.

The magnitude κ must be positive if the system is to give only imaginary values for h.

It follows from Equations (10), that in any arbitrary small rotation of the system:

$$\frac{1}{\kappa} XdX + YdY + ZdZ = 0.$$

Therefore:

$$X^2 + \kappa Y^2 + \kappa Z^2 = \text{const.}$$

So if we express X, Y, Z in terms of dr, ds, dt by means of Equations (6b) and (2a), and put:

$$dS^2 = (l_0 dr + m_0 ds + n_0 dt)^2 + \\
+ \kappa(\lambda_0 dr + \mu_0 ds + v_0 dt)^2 + \\
+ \kappa(\lambda_1 dr + \mu_1 ds + v_1 dt)^2$$

it then follows that ds is a magnitude which remains constant under all rotations of the system about the point $dr = ds = dt = 0$, and is of the same order of small magnitudes as dr, ds, and dt themselves [27].

Thus this magnitude can be used as a measure, independent of rotary motions, of the spatial difference between the points (r, s, t) and $(r + dr, s + ds, t + dt)$.

4

With this we have got to the starting point of Riemann's investigations. For it has emerged that there exists a homogeneous second degree expression of the differentials, which remains unchanged during any motion of two points which are fixedly connected to each other and whose separation is vanishingly small [28].

In doing this, we have applied the axioms II to IV laid down above, which express the possibility of congruence between different parts of space, to infinitely small spatial elements alone. Riemann's assumption, it therefore emerges, is identical with the assumption that space is monodromic, and that infinitely small spatial elements are in general congruent to one another, disregarding their limits. The sense of this proposition becomes intuitive when one restricts it to two dimensions [29]. It follows from Riemann's assumption in this case, that spatial measurements are the same as those which our analytic geometry instructs us to perform on an arbitrarily curved surface. Indeed, the infinitely small surface elements of an arbitrary curved surface are all to be regarded as flat, and therefore as congruent to one another if one disregards their limits.

The rest of the investigation then concerns the consequences, which result, if one demands congruence, independent of limits and under all possible rotations yielded by postulate III, between the finite parts of space. For this case in two dimensions, the curved surface must turn into a spherical surface*, or into a surface which arises from a spherical

---

* (1882) or a pseudosphere. The possibility of pseudospherical geometry is overlooked

one by bending without stretching. Similarly, Riemann has shown for three or more dimensions, that the magnitude called by him the measure of curvature must be constant. This part of my investigation, which is contained implicitly in Riemann's, I will not develop further here. The result is as follows:

If our assumptions I to IV are satisfied, then the most general system of geometry is what the rules of our customary analytic geometry would give if one applied this to a three-dimensional spherelike* structure, whose equation would be expressed in terms of four rectangular coordinates X, Y, Z, S, as:

$$X^2 + Y^2 + Z^2 + (S + R)^2 = R^2.$$

In this X, Y, Z cannot become infinite unless $R = \infty$. The latter special case corresponds to our actual geometry conforming with the axioms of Euclid. X, Y and Z can then only have finite values if $S = 0$, which is the equation of a flat structure. In this sense we must, like Riemann, refer to Euclid's space, in relation to spaces having a higher number of dimensions, as flat space [30].

I will finally note again, that if one does not lay down postulate IV, there result wholly different systems of geometry, which yet nonetheless could be implemented consistently. This is most easily to be seen for two coordinates. If the magnitude θ of Equation (5b) were not zero, the linear dimensions of any plane figure would increase in constant proportion during rotation through a constant angle in the same direction. The line of physically equivalent distance from a point would be the spiral.

One obtains another example that can be treated easily, if in the analytic geometry of the plane, with rectangular coordinates, one takes y to be imaginary [31]. That corresponds to the case where h_1 and h_2 are real and

$$h_1 + h_2 = 0.$$

The line of equivalent distance from a fixed point would then be an equilateral hyperbola [32].

here, as was noted above [i.e. on p. 617 of the second volume of his *Wissenschaftliche Abhandlungen* – P.H. & M.S.]
* or pseudospherical.

Taken together, Riemann's and my investigations therefore show, that the postulates laid down above when combined with the two propositions:

　　V.　that space has three dimensions
　　VI.　that space is infinitely extended

yield the foundation which suffices for developing the theory of space*. I have already emphasized that the correctness of these same postulates also has to be presupposed in customary geometry, if indeed silently. Our postulates therefore assume less than is presupposed by the geometrical proofs which are customarily given.

At the same time, I call attention to the fact that, as it clearly emerges in this development, the whole possibility of the system of our spatial measurements depends upon the existence of natural bodies corresponding sufficiently closely to the concept of fixed bodies laid down by us. The independence of congruence from place, from the orientation of the coinciding spatial structures, and from the path along which they were led to each other, is the fact upon which the measurability of space is based.

NOTES AND COMMENTS

[1] Helmholtz' views have been criticised and augmented by S. Lie**, and portrayed by him with the resources of group theory. It turned out that in order to be able to preserve Helmholtz' calculations one must express his axioms a little differently (note 15). They also contain dispensable components (note 24).

As Helmholtz, Lie considered motions of spaces (of rigid bodies). The new coordinates of a moving (material) point thereby become functions of the old ones. But Lie had to presuppose that these functions are differentiable. The group-theoretic axiomatic presentation was freed by Hilbert*** from both this and yet another restriction, though with his limiting himself to the planimetric case. He presupposes essentially the following:

Two successive continuous transformations of the plane (thus differentiability not demanded) yield another one (the motions form a group). A rotation (motion with a point held fixed) takes a moving point into infinitely many others. To this there is added a further continuity requirement.

Recently, H. Weyl (*Raum, Zeit, Materie* ['Space, time, matter'], 4th ed., Berlin, 1921,

---

* However, they do not yet distinguish between Euclid and Lobatchewsky.
** S. Lie, *Theorie der Transformationsgruppen* ['Theory of transformation groups'], Leipzig, 1873, vol. 3, pp. 437–523.
*** *Grundlagen der Geometrie* ['Foundations of geometry'], 4th ed., Leipzig, 1913, Appendix 4, p. 163.

p. 124) has given group-theoretic axioms for manifolds far more general than Euclidean or Riemannian ones.

[2] This sentence indicates the difficulty on account of which any empiricist theory of the origin of 'physical' geometry is threatened with getting caught in a circle. The latter is in fact only avoidable if we assume that some or other propositions, which are not judgments of experience, underlie geometry. Riehl therefore thinks (*Helmholtz in seinem Verhältnis zu Kant* [op. cit.], p. 32; *Kantstudien* **9** (1904), 276) that Helmholtz by this statement "conceded everything that Kant actually claimed in asserting that geometry is *a priori*". Yet such propositions could very well play, in 'physical' geometry, the role of definitions. Not indeed definitions in the customary sense of the word – Helmholtz would then also not be allowed to speak of examining their 'factual correctness' – but they could together form a system of implicit definitions (see e.g. M. Schlick, *Allgemeine Erkenntnislehre* [op. cit.], §7), which would as a whole be accessible to empirical examination.

[3] B. Riemann, *Über die Hypothesen, welche der Geometrie zugrunde liegen* ['On the hypotheses underlying geometry'], in *Werke*, 2nd ed., 1892, p. 272; new edition by H. Weyl, Berlin, 1919.

[4] As is known, Helmholtz' axiom system was not the first to be laid down. Thus one is required to examine what could justify a new treatment of the foundations. We wish to deal with this question here more fully than does Helmholtz himself.

When phenomena are connected by a system of propositions, it suffices to place some of them at the head as axioms. The others can then be derived from these. The selection of axioms can however take place according to various criteria:

(1) Those propositions can be made primary which are not merely logical grounds for the others, but also 'objective' ones (in the case of fields for which this distinction has a sense). Attention has been drawn to the significance of this concept especially by B. Bolzano[*], following Leibniz. Logical grounds for the law of mass action are given in e.g. the laws of thermodynamics, but objective grounds – according to the kinetic theory of gases – in the laws of mechanics. The concept of an objective ground can be discussed more precisely only in connection with the concept of a cause.

(2) The axioms to be placed at the head are to be so chosen that as few as possible are required. That many propositions can be embodied by a smaller number is possible because:

(a) from propositions as premises others logically follow;

(b) the functional form allows one to unite individual propositions into a *single* more comprehensive one.

(3) One will wish to start from elementary laws (differential laws). Requirement 2 is then satisfied as well. The fact that such an axiom system embodies fewer individual relationships is easily overlooked, because the laws thereby saved are in the other formulation united in functional form with other laws (see above under 2b)[†].

---

[*] B. Bolzano, *Wissenschaftslehre* ['The theory of science'], Sulzbach, 1837, vol. 2, pp. 339–350.

[†] [P. H. seems to think that $dy/dx = 2$ is more economical than $y = 2x$, because the latter is supposedly an abbreviation for $y_1 = 2x_1$, $y_2 = 2x_2$, etc. where x_1, y_1, etc. are the individual values of x and y. But it should rather be construed as the single sentence (x) (y) $(y = 2x)$ with universal quantifiers.]

(4) It has so far been presupposed that the system of propositions is given to us. The question may concern us of how we arrived at it, but we shall then put those propositions at the head which offer themselves first (immediately) in experience.

Before we apply these considerations to our problem, let us ask ourselves whether geometrical relations can relate according to considerations rooted in reality[†]. A geometrical relation, e.g. the extension a being equal to the extension b, manifests itself in a causal interrelationship between coincidences (see note 9). But we shall regard the relation itself as something different from causal interrelationships. Even when it is a member of a purely geometrical interrelationship (e.g. from $a=b$, $b=c$ there follows $a=c$), it does not occur in a causal interrelationship, although such a relationship (a ground of being, as Schopenhauer[*] puts it) is closely related to that of a real ground. Therefore the concept of an objective ground too probably cannot be applied to the relationship of geometrical propositions to each other.

Thus there remain, apart from the self-evident second requirement, only these two:

(1) to lay down elementary laws (differential laws);
(2) to lay down such laws which offer themselves first (immediately) in experience.

Now what does the system of propositions called by us geometry refer to? In part to the metrical relationships which occur in external things, in part to those of our mental images[††]. Here we only want, in the first instance, to consider geometry in the former respect. But even then metrical relationships, equality, etc. can be something to be discovered immediately in what is given, or something established by measurement. We want in the first instance to choose the latter viewpoint.

Riemann then satisfies the first requirement laid down above: he gives an elementary law (differential law). But this cannot be obtained immediately in experience. Helmholtz conversely formulates his axioms for systems of finitely separated points[**], because only about these can one assert something experimentally. In this there emerges the empiricist tendency of Helmholtz' undertaking, and also in another point:

One usually starts with the concept of equality as one incapable of much reduction. In fact one can appeal to immediate intuition. To this are then added propositions which establish the meaning of the concept for physical events. If one only wishes to deal with the results of measurements, one can initially forgo an explanation of the basic concepts, develop the geometrical system (so that the basic concepts are therefore given by implicit definitions: M. Schlick, *Allgemeine Erkenntnislehre* [*op. cit.*], §7), and then add physics. In this way the basic geometrical concepts are then indirectly defined by certain physical phenomena. As opposed to that, Helmholtz from the start fills the places of several of these concepts with certain experimentally ascertainable states of affairs.

If the basic geometrical concepts (e.g. equality of extensions) are moreover to have an intuitive meaning, then as regards understanding too Helmholtz' system seems to make smaller demands up on one's power of representation, because we have to represent to ourselves only points and not straight lines or plane surfaces.

[†] ['Realgrundverhältnis'; 'real cause' would be more appropriate, had P.H. not just distinguished between 'ground' and 'cause'.]

[*] Über die vierfache Wurzel des Satzes vom zureichenden Grunde ['On the fourfold root of the principle of sufficient reason'], §35. [Schopenhauer's term is 'Grund des Seins'.]

[††] ['unserer Vorstellungsbilder']

[**] Admittedly, his calculations can only be carried out if the axioms are asserted for infinitesimally neighbouring points, as S. Lie has proved (see note 15).

On the other hand, Euclid's axioms have the advantage that their truth is evident immediately in the intuition of the simultaneously given, whereas with Helmholtz' axioms even understanding is not possible other than in successive representations. Thus they can compel our assent at most on an associative basis.

Finally, from the viewpoint of relativity theory the invariant character of Helmholtz' axioms is noteworthy, inasmuch as they can be formulated without reference to a coordinate system.

[5] See Riemann's *Werke*, 2nd ed., Leipzig, 1892, p. 274; edition of H. Weyl, Berlin, 1919, p. 3. See also notes II.8 and III.29 of this volume.

[6] This requirement has been dropped by Weyl, as a vestige of prejudice about geometry 'at a distance'. See note 11.

[7] In presupposing the possibility that the coordinates of a point may change, material points are distinguished from spatial points. That presupposes that there is a sense in talking of the *same* material point. But for this one needs to associate with each other certain states, observed at various times, as states of 'the same material particle'. One does not, however, thereby need also to ascribe to the material particle a separate existence distinct from these states, and the association is not objectively given, but devised according to principles of economy of thought, and indeed ordinarily includes the use of continuity properties.

The theory of matter has been worked upon above all by G. Mie. It sets itself the task of laying down equations between the state magnitudes such that there arises from them, in an unforced manner, the existence of certain regions distinguished by the state magnitudes, which can be spoken of as substance. So far, however, one has not succeeded in explaining in this sense the existence of the electron, or of its counterpart the hydrogen nucleus.

In order, secondly, that there should be coordinates, there must exist three kinds of thing:
(1) the body whose coordinates are to be specified;
(2) a fixed body, to which the coordinates are to refer (reference system);
(3) measuring rods, i.e. rigid bodies which are brought into contact variously with points of the coordinate system and ones of the given body.

Both the coordinate systems and the measuring rods must be rigid. What a rigid body is, however, Helmholtz defines in precisely the following study, and thereby making use of the concept of coordinates. There seems to be a circle here. Such a circle is avoided if one picks out by definitions (which refer to a genus, not by individually displaying the objects concerned – see note III.12) a group of bodies (e.g. bodies at very low temperatures) having the property that when some of them are used as coordinate systems and measuring rods, the others thus measured can count as rigid*. One can at any rate thus define the concept of rigidity. If one however wants to explain what counted, in the old Newtonian mechanics, as a coordinate in absolute space, further supplementary considerations are required. But even if, as has been done here, we explain the concept of a coordinate without their help, all of Helmholtz' subsequent account can be carried through.

It seems, however, plausible to think equally of the stationary reference space as a rigid

---

* Compare note I.31. One can adopt similar considerations in order to define the concepts of temperature and ideal gas.

body. One could call coordinates of the moving point numbers corresponding to those coordinates, *once* measured, of the points of the immobile body, with which the moving point variously coincides. While even the use of the measuring rod could finally be wholly dispensed with, by associating some or other kind of metrical numbers (curvilinear coordinates) with the points of the stationary rigid body. Yet three-dimensional rigid bodies in three-dimensional space are impenetrable. Thus that stationary space is only an object of thought, which the rigid body can only be referred to by a fresh application of the measuring rods each time. It is a different matter in the planimetric case – for one can let one lamina slide along another.

Analytically, the coordinates will be regarded as a function of time and of parameters characterising the material points of the fixed body. S. Lie* has drawn attention to the fact that Helmholtz presupposes the existence of the second differential of these functions (see also note 1).

[8] If one chooses three arbitrary colours as basic colours, then starting from them one can portray any other colour (in a domain). Consequently, one can associate with any colour the point whose coordinates are equal to the three quantities used of the basic colours in portraying it**.

By mixing two colours there arises out of them a specific third one. If one therefore also associates with each colour the vector from the origin to the point constructed as above, the following holds: the vector of the mixed colour is the vector sum of the colours which combine. Other geometrical operations are initially (though see the end of this note) not defined. A geometry in which only the vector sum has a meaning independent of the coordinate system, is called affine. We can therefore also say that the laws of colour mixing obey the laws of affine geometry.

Should we choose different basic colours, but the same notation space, then the points standing for a given complex of colours will shift. The transformation is an affine one, producible by rotation and expansion. Now the preceding discussion in the text concerned an actual motion of material points, not an alteration in the reference system; thus the comparison made by Helmholtz does not simply refer to exactly corresponding situations. All the same, the system of points corresponding to one and the same set of colours is subject, when the choice of basic colours is changed (and the coordinate axes of the symbolic space retained), to an affine transformation.

Yet there exist three equations already between *four* points. If e.g. the colour D is the sum of the colours A, B, C, then:

$$x_D = x_A + x_B + x_C, \qquad y_D = y_A + y_B + y_C, \qquad z_D = z_A + z_B + z_C,$$

which holds quite independent of the choice of basic colours. Thus we should probably understand by the preceding passage, that Helmholtz further wants to give all possible situations to the coordinate system of the notation space (especially to its origin) in which the points are set. The affine transformations which then enter into consideration are characterised by 12 coefficients:

---

* *Theorie der Transformationsgruppen* [*op. cit.*], vol. 3, p. 440.
** See e.g. Helmholtz, *Physiologische Optik* ['Physiological optics'], 2nd ed., 1892, pp. 326ff.

$$x' = \alpha x + \beta y + \gamma z + a$$
$$y' = \delta x + \varepsilon y + \zeta z + b$$
$$z' = \eta x + \theta y + \iota z + c.$$

The 15 coordinates of the point quintuples arising from a given one by affine transformation, are therefore given by 15 equations with 12 arbitrary constants, so that after eliminating the constants three equations are left over. To this there corresponds the arbitrary expandibility of bodies in affine geometry.

Helmholtz* himself tried to extend the affine geometry of colours to a metric one; these attempts have recently been continued by E. Schrödinger**.

[9] It remains to be shown that each of these equations successively obtained is independent of the preceding one(s), thus that each one added truly implies a new restriction. The relevant proofs have been given by S. Lie***.

So we note the following: in a movement of the rigid body each point in the space is associated with another one, namely the one later occupied by the same material point as earlier occupied the given one. An association whereby each point in a space is given another one to correspond to it, is called a transformation. The considerations in the text then show that there exists a system of $n(n+1)/2$-fold infinitely many transformations.

But this system could still depend on the choice of rigid body and its initial situation. However, by adducing axiom II one recognises that this is not the case. For if at one time the material points A, B and at another time A', B' (with a change of situation of the rigid body), can be brought into coincidence with the spatial points a, b, then A', B' and A, B are congruent. If A, B can moreover coincide with a', b' too, then the same must also hold, according to axioms II and III, for A', B'.

A transformation of not more than $n(n+1)/2$ parameters thus truly exists *independent* of the choice of fixed body and its initial situation. Since therefore the displaced body only yields us transformations from the system of those already defined, it follows that two transformations from the system, performed successively, yield a transformation itself belonging to the system. The totality of these transformations is a *group*****.

[10] That is obviously supposed to mean: two rigid bodies.

[11] One will understand the sense of these statements even better, if one considers cases in which the behaviour described here does not occur. Firstly then, it might be that two originally coinciding rigid bodies could no longer be brought into coincidence if they were brought together again *along different paths*. We want to go into this later; for the time being, we want to think of the two bodies as being only taken along the same path. It will be pronounced selfevident that if bodies constituted alike and moved together once coincide, they will continue to do so. Yet let us think of the example on p. 7 of the portion of surface on the egg-shaped surface. We thus see that conceivably certain (rigid) bodies might not be movable at all. But the third axiom says that this case cannot in fact occur.

---

* *Zeitschr. f. Psychologie u. Physiologie* **2** (1891), 1; **3** (1891), 1, 517; *Berliner Berichte*, 1891, p. 1071; *Wissensch. Abhandlungen*, vol. 3, pp. 407–475.
** *Ann. d. Phys.* **63** (1920), 397, 481.
*** *op. cit.* pp. 446f., p. 523.
**** Lie, *op. cit.*, p. 449.

It is moreover useful to go into the processes in curved space a little further. Actually existing bodies can only count as rigid in the face of small forces; in the face of large ones they would break up, or else stretch or contract, as occurs with a surface portion made of suitable material which is moved on an egg-shaped surface. Yet how could we ascertain the stretching in such a case? Two even not quite rigid bodies constituted alike and moved along the same curve will after all, if they once coincide, permanently remain coincident. Shall one not say that, by definition, the new shape of the body is congruent with the old one? Obviously not, for the metrical relationships are transmitted by *infinitesimal* measuring rods. Thus in the presupposed case too the curvature of the space will be ascertainable from motions (even when only ones along one and the same curve are considered). If e.g. before the displacement an (infinitesimal) portion of body can be brought into coincidence with a *part* of another finite portion, then this might no longer be the case after the displacement. We therefore have to emphasize, as an especially important feature of Helmholtz' axiom, the following proposition: two point systems which once coincide can also be brought into coincidence in every other situation, even when each is connected to another system.

As is known, Riemann considered a more general case: finite rigid bodies need not always be movable, but infinitely small bodies should be able to go anywhere; thus, in other words, the axioms stated here by Helmholtz should hold for infinitesimal rigid bodies: there should be a line element independent of the path. In this Weyl* sees a vestige of prejudices about geometry "at a distance". The length relationship between two extensions could depend upon the path along which an infinitesimal comparing rod was brought from one to the other. H. Reichenbach** can already see a further generalisation: that namely a material extension, after a rotation about itself, might no longer coincide with the same extension as previously. (Invalidity of Helmholtz' monodromic principle, see p. 44.)

Let us now look at things in terms of Riemann's assumption. Does the equation of separation for two pairs of neighbouring points then contain the ground for the described interrelationships between them, or *is* it nothing other than the law of these interrelationships? It is perhaps only a question of a different form of expression. But should one choose the second form, one gets laws of a quite peculiar logical structure***. A reduction without remainder of geometry to causal interrelationships would also hardly succeed in this manner. It may well be simplest to regard metrical properties as ultimate realities.

¹² A fixed body rotatable about a point possesses three degrees of freedom, since its situation is determined by the orientation of an axis fixed in it, and by the rotation of the body about this axis. One would thereby expect that the retinal image of one and the same body, as the position of the eyeball varies, can take up threefold infinitely many situations, as also a portion of surface on a sphere and on a plane may have threefold infinitely many situations. It follows from the laws of Donders and Listing**** that this is not correct. When the line of sight is given, the position of the eyeball is also determined and is thus no longer capable of any rotation.

---

* See e.g. *Raum, Zeit, Materie* [*op. cit.*], §16; *Math. Zeitschr.* **2** (1918).
** *Relativitätstheorie und Erkenntnis a priori* ['Relativity theory and *a priori* knowledge'], Berlin, 1921, p. 76.
*** If it in particular holds that if *a* then *b*, then if *c* also *d*.
**** Helmholtz, *Physiologische Optik* [*op. cit.*], pp. 619 and 621.

[13] Helmholtz, *Physiologische Optik* [*op. cit.*], pp. 690 f.

[14] See note 24.

[15] One is not allowed to apply to the infinitely small the axioms which were laid down for the finite case, as S. Lie has shown.

If one point is held fixed, then an arbitrary point can still occupy twofold infinitely many situations. That is indeed still correct if the point concerned gets very near to the one held fixed. Yet it might be that certain situations were no longer distinguishable, i.e. would involve only differences of coordinates higher than the first order of the infinitely small. Then e.g. an observer located very near to the fixed point would only ascertain *one* degree of freedom of mobility for the other points.

Lie (*op. cit.*, p. 456) has illustrated this with an example[†]. If x, y, z denote the coordinates of a point before a displacement and $\bar{x}$, $\bar{y}$, $\bar{z}$ those after one, then let:

$$\bar{x} = x + \alpha$$
$$\bar{y} = y + \beta + \gamma x + \delta x^2 + \varepsilon x^3 + \zeta x^4 = y + \phi^4(x)$$
$$\bar{z} = z + \gamma + 2\delta x + 3\varepsilon x^2 + 4\zeta x^3 = z + \phi^{4\prime}(x)$$

where $\phi^4(x)$ is a fourth degree function and $\phi^{4\prime}(x)$ is its derivative.

One can still express the transformation in the same manner after a translation of the coordinate system:

$$\bar{x} - A = x - A + \alpha\,; \qquad\qquad \bar{x} = x + \alpha$$
$$\bar{y} - B = y - B + \phi^4(x - A)\,; \qquad \bar{y} = y + \psi^4(x)$$
$$\bar{z} - C = z - C + \phi^{4\prime}(x - A)\,; \qquad \bar{z} = z + \psi^{4\prime}(x).$$

If one carries out two transformations successively, one gets:

$$\bar{\bar{x}} = \bar{x} + \bar{\alpha} = x + \alpha + \bar{\alpha}$$
$$\bar{\bar{y}} = \bar{y} + \overline{\phi^4}(\bar{x}) = y + \phi^4(x) + \overline{\phi^4}(x + \alpha) = y + \chi^4(x)$$
$$\bar{\bar{z}} = \bar{z} + \overline{\phi^{4\prime}}(\bar{x}) = z + \phi^{4\prime}(x) + \overline{\phi^{4\prime}}(x + \alpha) = z + \chi^{4\prime}(x)$$

Thus the transformations form a group.

Now let one point be held fixed, we shall make it the origin. Our transformation equations then read:

$$\bar{x} = x$$
$$\bar{y} = y + \delta x^2 + \varepsilon x^3 + \zeta x^4$$
$$\bar{z} = z + 2\delta x + 3\varepsilon x^2 + 4\zeta x^3.$$

There would thus be ∞^3 different situations of the body, and a point could still, since its x-coordinate remains constant, occupy twice ∞ many situations.

But if infinitely small terms of higher order are neglected, one can write:

[†] [This example uses the analysis of note 9 above in reverse. A group of transformations, of the type obtained there, is defined such that all bodies whose range of possible motions is given by the group are rigid in the sense of axioms II and III. Thus the inclusion of all translations in the group corresponds to the absolute mobility, as demanded by axiom III, of the first point selected in the body.]

$$\bar{x} = x$$
$$\bar{y} = y$$
$$\bar{z} = z + 2\delta x.$$

Thus if one only looks at the immediate vicinity [of the point held fixed], one will only find ∞^1 different situations of the body. Each point, since its x and y-coordinates remain constant, can only occupy ∞^1 many situations, and every line element through the point held fixed can only move in a plane going through the z-axis.

These difficulties are avoided if one asserts Helmholtz' axioms directly for the infinitely small case (see Lie, *op. cit.*, pp. 460ff.).

[16] Concerning the concept of coordinates, see note 7. The u, v, w therefore characterise the material point. In laying down the so-called Lagrange equations (which derive from Euler) of hydrodynamics, one likewise individualises each liquid particle separately by citing their places at a set time*.

In hydrodynamics this procedure is customary because the liquid mass constantly changes its form. It is simpler here to think of the measurement performed directly on the fixed body. But even that is superfluous; one can continuously associate, in an arbitrary manner, with each material point a value triple.

[17] Thus what are being considered are rotations of the rigid body about a point in it.

[18] The new notation is probably introduced in order to avoid the differential symbol with this meaning. It will namely later (see note 20) be used with another meaning.

One can either understand by εx, εy, εz the coordinates of a neighbouring point, or by x, y, z three magnitudes which behave as the coordinates of arbitrary points on the tangent of a curve going through the origin; in this sense one can also characterise the three values as coordinates of the line element. There then also arise two points of view by which to understand everything that follows:

(1) Where x, y, z are spoken of in what follows, the coordinates of the line element are intended. Only the proportions $x:y:z$ possess a meaning.

(2) x, y, z means a point.

Naturally, one cannot in the strict sense immediately associate a point with the values x, y, z. That can only be done in terms of limit considerations. Let us e.g. speak of a rotation about the point 0, 0, 0, with a second point x_0, y_0, z_0 held fixed. What is intended by that? Instead of the *one* rotation about 0, 0, 0 and x_0, y_0, z_0, let us consider a series of cases F_ε. In case F_ε let the point εx_0, εy_0, εz_0 be held fixed as a second point, and the figure then be enlarged $1/\varepsilon$ times from 0, 0, 0 outwards. One then obtains image processes for all of which the points 0, 0, 0 and x_0, y_0, z_0 are fixed points. The trajectory curves in these image rotations will converge towards limiting curves, and let us understand by x', y', z' always a point lying on the limiting curve which goes out from x, y, z.

Correspondingly, there are also two possible ways of interpreting the fourth axiom, whereby this acquires different ranges of significance:

(1) If one rotates the body about a line element $x:y:z$, then at some time a line element x, y, z must go over into itself again, in other words**:

---

* See e.g. Riemann-Weber, *Partielle Differentialgleichungen der Physik* ['Partial differential equations of physics'], 5th ed., Brunswick, 1912, vol. 2, p. 411.

** Compare Lie, *op. cit.*, p. 461.

$$x':y':z'=x:y:z.$$

One can also say that one must have (with notation as in the text):

$$\frac{d\rho}{dr}=\frac{d\sigma}{ds}=\frac{d\tau}{dt}$$

(2) If one understands, as we explained under 2 above, by x', y', z' values lying on the trajectory curve going through x, y, z, then at some time one must have:

$$x'=x; \qquad y'=y; \qquad z'=z.$$

One can also avoid the limit consideration and demand that at some time once again:

$$\frac{d\rho}{dr}=\frac{d\sigma}{ds}=\frac{d\tau}{dt}=1.$$

Naturally, more is demanded here than is in case 1. The question arises of whether perhaps the requirement of case 1 is sufficient.

[19] Against this there are the reservations advanced in note 15.

[20] Here we now have the other meaning (see note 18) of the differential symbol, which from now on is the only meaning entering into consideration. By dr was meant the increase of the r-coordinate from one point to a neighbouring one; as opposed to this, dx, dy, dz give the change of the orientations of a line element which is fixedly connected to a material point, when one makes the transition from one situation of the fixed body to a neighbouring one. This transition can be taken to occur in time, and one will then characterise as components of velocity the differentials dx, dy, dz divided by the temporal duration of the transition.

If one makes the transition from the old situation of the fixed body to the new one, the coordinates belonging to the same material point then change by small values. In assigning to a totality of numerical values small changes, such that these changes depend upon the numerical values, one performs in Lie's terminology an *infinitesimal transformation*. The infinitesimal transformations arising from the motion of a fixed body form a *group*, i.e. by putting two together there arises a new one. In what follows one will be looking for the non-infinitesimal group to which the infinitesimal one belongs.

[21] We can now take the third situation (p. 45) ρ, σ, τ as fixed and the second as variable. Then ρ, σ, τ and ξ, υ, ζ will characterise the material point, while r, s, t and x, y, z characterise the coordinates. The difference between (3a) and (3b) is that the former gives the velocity as a function of the material point, the latter as a function of the location in space. We need (3b) to continue further.

It might be thought initially that if the location with the given x, y, z has been occupied by another material point, then a system of equations with the *same* a_0, b_0, c_0 ... would no longer be possible. But such an assumption is forbidden by axiom II. According to the latter, the displacement expressed in terms of x, y, z which was possible before the change of location must also be possible after the change of location. In particular, we can put together transformations with always the same a_0, b_0, c_0..., i.e. integrate (3b). But (3a) cannot be handled thus (see pp. 48–49).

[22] If these equations are to hold for all x_0, y_0, z_0, then we must have $a_0=\cdots=c_0=0$. Since however only certain points stay put during a rotation, the sentence only starts to be correct

if the words "arbitrarily given values of x_0, y_0, z_0" are replaced with the words "certain values of x_0, y_0, z_0".

But even this much cannot yet be asserted at this point. We do indeed know that in every rotation of Euclidean space a straight line stays put. However, this state of affairs cannot be immediately inferred from axiom III, but rather only the converse – that to every axis there belongs a rotation.

For this reason one must replace the formulation of the text as follows: "Therefore, for an arbitrary x_0, y_0, z_0 we must be able to choose the quantities dp', dp'' and dp''' such that there holds for the corresponding a, b, c:

$$0 = a_0 x_0 + b_0 y_0 + c_0 z_0$$
$$0 = a_1 x_0 + b_1 y_0 + c_1 z_0$$
$$0 = a_2 x_0 + b_2{}_0 + c_2 z_0$$

from which there follows:

$$\begin{vmatrix} a_0 & b_0 & c_0 \\ a_1 & b_1 & c_1 \\ a_2 & b_2 & c_2 \end{vmatrix} = 0.\text{"}$$

This formulation however also suffices for what follows.

The following sentence "This can only occur if ... (4)" is nonetheless correct if (as Professor Engel has commented by letter) one understands by rotations not motions with *one* point held fixed, but ones with *two* held fixed. We have already commented that it does not follow directly from the axioms that rotation means the same thing in both senses.

[23] One can also make the substitution:

$$x = L e^{hn}, \qquad y = M e^{hn}, \qquad z = N e^{hn},$$

when one obtains the system of equations:

$$Lh = La_0 + Mb_0 + Nc_0$$
$$Mh = La_1 + Mb_1 + Nc_1$$
$$Nh = La_2 + Mb_2 + Nc_2.$$

The corresponding determinant:

$$\begin{vmatrix} a_0 - h & b_0 & c_0 \\ a_1 & b_1 - h & c_1 \\ a_2 & b_2 & c_2 - h \end{vmatrix}$$

arises from (4b) by reflection through the principal diagonal.

[24] This equation means: there is a bunch of invariant planes which go over into themselves during the rotation. After introducing the Euclidean metric, one recognises in them the planes perpendicular to the axis of rotation.

We may understand by x, y, z the coordinates of the line element. Then only the proportions $x : y : z$ have a meaning, and only 0 should stand on the right hand side of Equation (5a)*.

---

* This was pointed out to the editors by Professor Engel.

We saw, however, (note 18) that another interpretation is also possible: by x, y, z one may actually understand neighbouring points. Equation (5a) then means: during the rotation the end point of a vector is restricted to a plane. From this it follows that when the vector reaches its old orientation, its magnitude is the same as before. So if above (note 18) we distinguished two interpretations of axiom IV, we now see that the second formulation requires too much. It suffices in fact to lay down the first requirement*.

However, the monodromic principle is wholly dispensable, as Lie has shown. It is only necessary to presuppose that the body can still rotate after holding a line element fixed, but not if one also holds fixed besides this a surface element going through the line element**. This condition naturally only refers to the case of geometries of three or more dimensions. In setting up two-dimensional geometry the monodromic principle is indispensable (see Helmholtz, this edition p. 57).

[25] These are already the formulae for the rotation of a rigid body about the X-axis in the Euclidean space X, Y, Z.

[26] Thus, in the Euclidean interpretation, a rotation about an axis which is perpendicular to the one previously chosen.

[27] The function dS^2 is obviously determined only down to an arbitrary constant. The coordinates could indeed initially have been chosen quite differently.

[28] A standardisation (Weyl, *op. cit.*, p. 109) is still required, in order for the dS^2 to be the condition of transferibility for distant line elements too (see the previous note). That is however possible by axiom II.

[29] Curvilinear coordinates were introduced by Gauss***. One can denote each point on a surface with a pair of values p, q. In this way the surface is covered with two systems of curves, $p =$ const. and $q =$ const., so that each point is also characterised as the intersection of two curves, *one* from each system. The square of the line element, as measured in ordinary three-dimensional Euclidean space, is then expressible as a quadratic form of dp and dq, the differentials of the coordinates:

$$ds^2 = E\,dp^2 + 2F\,dp\,dq + G\,dq^2,$$

where E, F, G are functions of p and q. However, the two-dimensional manifold of the p and q is now by no means Euclidean, but precisely a manifold of the more general kind considered here.

In this sense one can now also equate two-dimensional Bolyai-Lobatschewsky geometry with the geometry on a certain surface of Euclidean three-dimensional space (see Helmholtz, this edition p. 8), and so-called spherical geometry with the geometry of the spherical surface. But such an interpretation is by no means obligatory. We can impose a non-Euclidean metric on an arbitrary two-dimensional manifold by making corresponding

---

* Lie, *op. cit.*, p. 461.
** Lie, *op. cit.*, pp. 477, 481. The editors also owe the reference to this passage to the kindness of Professor Engel.
*** Disquisitiones generales circa superficies curvas ['General enquiries concerning curved surfaces'], §4. (In vol. 4 of Gauss' collected works, Göttingen, 1863–1871; originally published 1828.)

presuppositions about the square of the line element, or about the behaviour of rigid bodies (see note 1.25 and the corresponding part of Helmholtz' text).

Three-dimensional spherical space can be thought of as embedded in a flat four-dimensional one. But that too is not necessary. It is therefore a misunderstanding to suppose that the finitude of the world, surmised by Einstein, can only be thought of by presupposing a four-dimensional space.

[30] Various possibilities are overlooked[†]. We shall restrict ourselves to the two dimensional case.

(1) The space could be a two-dimensional structure of constant positive curvature, yet without being a spherical surface. To make this structure intuitive one need only take two symmetrically opposite points of an ordinary spherical surface to be a "point", and the great circles on the sphere to be straight lines. Then two "straight lines" always have *one* and only one point in common, and all axioms of ordinary geometry hold except the axiom od parallels (elliptical geometry, see F. Klein, *Werke*, vol. 1, pp. 249, 287, 401).

(2) The space can be a finite surface having everywhere the measure of curvative zero (a Clifford-Klein surface). One may take a plane lattice and call the totality of homologously situated points a "point". Or one may bend a rectangular piece of paper to form a cylinder, and then this – with stretching – to form a torus (ring), by bringing the two base surfaces of the previously produced cylinder into contact. However, let the old metric be retained on the torus, i.e. ascribe to a pair of neighbouring points the separation which they had before the stretching (Clifford, *Werke*, pp. 139f.; Klein, *Werke*, vol. 1, pp. 355, 369; Killing, *Grundlagen der Geometrie* ['Foundations of geometry'], Paderborn, 1893, vol. 1, p. 271).

[31] The element of the ar then assumes the form:

$$ds^2 = dx^2 + (idy)^2 = dx^2 - dy^2.$$

It is, as one says, 'indefinite'. The distance from the origin is $\sqrt{x^2 - y^2}$.

[32] If we namely use the notation[††] of (6b), then by (5) and (5a):

$$X' = X, \qquad Y' = Ye^{h_1\eta}, \qquad Z' = Ze^{-h_1\eta}$$

thus $Y'Z' = YZ$, or if we put

$$Y = \bar{Y} + \bar{Z}, \qquad Z = \bar{Y} - \bar{Z}$$

then:

$$(\bar{Z}')^2 - (\bar{Y}')^2 = \bar{Z}^2 - \bar{Y}^2.$$

If we only assume rotations about the X-axis to be possible, we can restrict consideration to points in the plane $X = 0$. For the sake of simplicity, we shall consider a two-dimensional fixed body.

Thus the points satisfying the equation

$$\bar{Z}^2 - \bar{Y}^2 = \text{const., e.g.} = 1$$

[†] [If the spaces discussed in this note are not in fact intuitively "imaginable" or "representable" in Helmholtz' sense, it is incorrect to say that they have been "overlooked". See notes I.35, I.36.]

[††] [In what follows, by X', Y', Z' is meant the point which follows X, Y, Z after the elapse of η.]

lie on a hyperbola. Since they can be transformed into one another, one ascribes to them the same distance from the origin, e.g. 1. As a general expression for the square of the distance from the origin, one takes $\bar{Z}^2 - \bar{Y}^2$, and as the square of the line element:

$$ds^2 = d\bar{Z}^2 - d\bar{Y}^2.$$

We will retain the notation of the preceding note, and put:

$$ds^2 = dx^2 - dy^2$$

and for the square of the distance: $x^2 - y^2$. Then every point on the straight lines $x = \pm y$, which we will call OQ and OQ', will have the distance 0 from the origin.

Let us impart a rotation to the rigid body during which a straight line G fixed in it goes from OA_1 (different from OQ and OQ') to OA_2. If the process is repeated several times, then G reaches OA_3, OA_4, etc., and one will call the angles A_2OA_1, A_3OA_2, etc. equal. If moreover P_1, P_2, P_3, etc. are points occupied by a material point on G in the latter's various situations, then we must have $OP_1 = OP_2 = OP_3$, etc.

The straight lines OQ and OQ' accordingly can never be crosses, since every point on them has the distance 0 from the origin. Thus surface-like beings in whose plane surface rigid bodies showed this behaviour, would say: even a continually rotated straight line cannot regain its initial situation (stiff-necked geometry), the monodromic principle is not satisfied.

The requirement recognised by F. Klein as the component which is also indispensable for geometries of more than two dimensions, is that in rotations every straight line should return once again to its old orientation.

The requirement of Lie mentioned above (note 24) excludes, for geometries of more than two dimensions, the case of an indefinite element of arc too. If namely this is the case, then the straight line OQ automatically stays put during all rotations of the body, so that one can hold fixed a surface element containing OQ and still rotate the body.

Geometries with an indefinite element of arc have recently become of special interest. If one uses t to denote time and x_1, x_2, x_3 to denote the spatial coordinates, then the equation

$$dt^2 - dx_1^2 - dx_2^2 - dx_3^2 = 0$$

characterises those space-time points which can receive light signals from a given point. Whether a space-time point has this property is independent of the coordinate system. Thus the equation

$$(dt')^2 - (dx_1')^2 - (dx_2')^2 - (dx_3')^2 = dt^2 - dx_1^2 - dx_2^2 - dx_3^2$$

is moreover the condition for two coordinate systems t', x_1', x_2', x_3' and t, x_1, x_2, x_3 to be of equal status. When two systems are of equal status, they cannot be conceptually distinguished by statements concerning the outcome of experiments (the special theory of relativity).

CHAPTER III

NUMBERING[†] AND MEASURING
FROM AN EPISTEMOLOGICAL VIEWPOINT

From: *Philosophische Aufsätze Eduard Zeller zu seinem fünfzig-
jährigen Doktorjubiläum gewidmet* ['Philosophical essays dedicated to Eduard Zeller on the occasion of the fiftieth anniversary
of his doctorate'], Leipzig, Fues' Verlag, 1887, pp. 17–52.
Reprinted in *Wissenschaftliche Abhandlungen*, vol. III, pp. 356–391.

Although numbering and measuring are the foundations of the most
fruitful, sure and exact scientific methods known to us at all, relatively
little work has been done on their epistemological foundations. On the
philosophical side, strict disciples of Kant, who adhere to his system
exactly as it developed historically amidst the views and knowledge of
his time, had of course to regard the axioms of arithmetic as propositions
given *a priori*, which narrow down the specification of the transcendental
intuition of time in the same sense as the axioms of geometry do the
intuition of space. Through this conception, the issue of a further foundation and derivation for these propositions was terminated in both
cases.

In earlier writings I endeavoured to show that the axioms of geometry
are not propositions given *a priori*, but that they are rather to be confirmed and refuted through experience. Here I emphasize once again
that this does not eliminate Kant's view of space as a transcendental
form of intuition; in my opinion this merely excludes just one unjustified
particular specification of his view, although one which has become most
fateful for the metaphysical endeavours of his successors.

It is then clear that if the empiricist theory – which I besides others
advocate – regards the axioms of geometry no longer as propositions
unprovable and without need of proof, it must also justify itself regarding
the origin of the axioms of arithmetic, which are correspondingly related
to the form of intuition of time [1].

Up to now, arithmeticians have placed the following propositions as
axioms[††] at the head of their deductions:

† ['Zählen' has been translated by 'numbering' rather than 'counting', because Helmholtz
appears to have its etymological connexion with 'Zahl' ('number') in mind.]
†† [In English these axioms are normally stated in terms of 'equals'. But Helmholtz will

AXIOM I. If two magnitudes are both alike with a third, they are alike amongst themselves.

AXIOM II. The *associative law of addition*, as it is termed by H. Grassmann

$$(a+b)+c=a+(b+c).$$

AXIOM III. The *commutative law of addition:*

$$a+b=b+a.$$

AXIOM IV. Like added to like gives like.

AXIOM V. Like added to unlike gives unlike.

Hermann and Robert Grassmann* have got further with this investigation than the other arithmeticians whose work is known to me, while at the same time pursuing philosophical viewpoints. In what follows, I shall have to go along their path throughout my presentation of the arithmetical deductions. Amongst other things, they reduce the two axioms II and III to a single one which we shall call Grassmann's Axiom, namely:

$$(a+b)+1=a+(b+1),$$

from which they derive both of the above more general propositions by means of the so-called $(n+1)$-proof. I hope to demonstrate, in what follows, that the correct basis for the theory of the addition of pure numbers has thereby indeed been obtained. Regarding the issue, however, of the objective application of arithmetic to physical magnitudes, there is added thereby† to the two concepts of *a magnitude* and of *alike in magnitude*, both of whose sense in the realm of facts remains unexplained, a third one as well – that of the *unit*. And at the same time, it seems to me an unnecessary restriction of the domain of validity of the propositions discovered, that one should from the outset treat physical magnitudes only as ones composed out of units.

be seen to allude to the etymological connexions between 'gleich' ('equal', 'alike'), 'vergleichen' ('compare', 'liken') and a host of other words. So we translate 'gleich' in most occurrences by 'alike' or 'like', to be understood in the sense 'exactly alike'.]
* Hermann Grassmann, *Die Ausdehnungslehre* ['The theory of extension'], 1st ed., Leipzig, 1844 and 2nd ed., 1878. Robert Grassmann, *Die Formenlehre oder Mathematik* ['The theory of forms or mathematics'], Stettin, 1872.
† [i.e. in Grassman's Axiom; in what follows 'magnitude' translates 'Grösse' and 'alike in magnitude' translates 'gleich gross'.]

Among more recent arithmeticians, E. Schröder* has also essentially attached himself to the Grassmann brothers, but in a few important discussions he has gone still deeper. As long as earlier arithmeticians habitually took the ultimate concept of number to be that of a cardinal number[†] of objects, they could not wholly free themselves from the laws of the behaviour of these objects, and they simply took it to be a fact that the cardinal number of a group of objects is ascertainable independent of the order in which they are numbered. To my knowledge, Mr. Schröder (*op. cit.*, p. 14) was the first to recognize that here a problem lies concealed: he also acknowledged – in my opinion justly – that there lies a task here for psychology[2], while on the other hand those empirical properties[3] should be defined which the objects must have in order to be enumerable[††].

Besides this there also are relevant discussions, in particular about the concept of a magnitude, to be found in Paul du Bois-Reymond's *Allgemeine Funktionslehre* ['General theory of functions'], Tübingen, 1882, part I, ch. 1, and in A. Elsas' book *Über die Psychophysik* ['On psychophysics'], Marburg, 1886, pp. 49ff. However, both of these books deal with more specific investigations and without discussing the complete foundations of arithmetic. Both believe that one may derive the concept of a magnitude from that of a line, the former in the empirical sense, the latter in the sense of strict Kantianism. I have already mentioned above and explained in previous writings what I have to object against the latter viewpoint. Paul du Bois-Reymond closes his investigation with a paradox according to which two opposed viewpoints, both of which lead to contradictions, are alike possible.

As the author just named is a most acute mathematician, who has sought with particular interest after the deepest foundations of his discipline, the final result at which he arrived has encouraged me all the more to set out my own thoughts on this same problem.

In order to characterize, briefly at the outset, the viewpoint which leads to simple logical derivations and the solution of the mentioned contradictions, the following may serve. I consider arithmetic, or the

---

* *Lehrbuch der Arithmetik und Algebra* ['Textbook of arithmetic and algebra'], Leipzig, 1873.
[†] ['cardinal number' translates 'Anzahl' .]
[††] ['zählbar']

theory of pure numbers, to be a method constructed upon purely psychological facts[4], which teaches the logical application of a symbolic system (i.e. of the numbers) having unlimited extent and an unlimited possibility of refinement. Arithmetic notably explores which different ways of combining these symbols (calculative operations) lead to the same final result. This teaches us, amongst other things, how to substitute simpler calculations even for extraordinarily complicated ones, indeed for ones which could not be completed in any finite time. Apart from thereby testing the internal logicality of our thought, such a procedure would admittedly be primarily a pure game of ingenuity with dreamt up objects – one which Paul du Bois-Reymond scornfully compares with the knight's move on the chess board – if it did not permit such extraordinarily useful applications. For by means of this symbolic system of the numbers we give descriptions of the relationships between real objects which, where applicable, can reach any required degree of exactitude, and by means of it in a large number of cases where natural bodies, governed by known laws of nature, meet or interact, the numerical values measuring the outcome are calculated in advance.

But then one must ask: what is the objective sense of our expressing relationships between real objects as magnitudes, by using denominate[†††] numbers; and under what conditions can we do this? The question resolves itself, as we shall find, into two simpler ones, namely:

(1) What is the objective sense of our declaring two objects to be *alike*[†] in a certain respect?

(2) What character must the physical connexion between two objects have, in order that we may regard likenable[††] attributes of these objects as *additively* combined, and consequently regard these attributes as *magnitudes* which can be expressed by using denominate numbers? We namely consider denominate numbers to be composed out of their parts, or as may be units, by addition.

1. THE LAWLIKE SERIES OF NUMBERS

Numbering is a procedure based upon our finding ourselves capable of retaining, in our memory, the sequence in which acts of consciousness

---

[†] ['gleich']
[††] ['gleich', 'vergleichbar']
[†††] [Strict translation of *benannt*; Kahl uses 'concrete'–Ed.]

successively occurred in time. We may consider numbers initially to be a series of arbitrarily chosen symbols, for which we fix only a certain kind of succession as the lawlike or – as it is commonly put – the "natural" one. Its being termed the[†] "natural" number series was probably connected merely with one specific application of numbering, namely the ascertaining of the *cardinal number* of given real things. As we throw these one after another onto the already numbered heap, the numbers follow one another by a natural process in their lawlike series. This has nothing to do with the sequence of number symbols; just as the symbols differ in different languages, their sequence too could be specified arbitrarily, so long as some or another specified sequence is immutably fixed as the normal or lawlike one. This sequence is in fact a norm or law given by human beings, our forefathers, who elaborated the language. I emphasize this distinction because the alleged "naturalness" of the number series is connected with an incomplete analysis of the concept of number. The mathematicians term this lawlike number series that of the *positive whole numbers*.

The number series is impressed upon our memory extraordinarily much more firmly than any other series, which doubtless rests upon its much more frequent repetition. This is why we also prefer to use it in order to establish, through association with it, the recollection of other sequences in our memory; that is, we use the numbers as *ordinal numbers*.

2. UNAMBIGUITY[††] OF THE SEQUENCE

In the number series, the processes of going forwards and going backwards are not equivalent but essentially different, as with the sequence of perceptions in time, whereas for lines – which exist in space permanently and without change in time – neither of the two possible directions of advance is distinguished as against the other.

In fact every present act of our consciousness, be it perception, feeling or volition, works together in it with the memory images of past acts, but not of future ones, which are still not yet available in our conscious-

---

† [or maybe: "The symbolization of the..." ("Die Bezeichnung der"); note reference to 'number symbols' below.]

†† ['Eindeutigkeit', which Helmholtz may be thinking of as opposed to 'Zweideutigkeit', meaning literally that there is only one possibility of interpretation *as opposed* to two.]

ness at all; and we are conscious of the present act as specifically different from the memory images which exist beside it. The present representation is thereby contrasted, in an opposition pertaining to the form of intuition of time, as the succeeding one to the preceding ones, a relationship which is irreversible and to which every representation entering our consciousness is necessarily subject. In this sense, orderly insertion in the time sequence[†] is the inescapable form of our inner intuition.

3. THE SENSE OF THE SYMBOLISM[††]

According to the foregoing discussion, each number is determined only by its position in the lawlike series. We attach the symbol *one* to that member of the sequence with which we begin. *Two* is the number which immediately follows upon one in the lawlike series[5], i.e. without interposition of another number. *Three* is the number which likewise follows immediately upon two, and so on[6].

There is no reason for interrupting this sequence anywhere, or for returning in it to a symbol already used previously. The decimal system[7] indeed makes it possible, through combining only ten different number symbols in a simple and easily understandable way, to continue the series indefinitely without ever repeating a number symbol[*].

We shall call the numbers which follow a given number in the lawlike series *higher* than that one, and those which precede it *lower*[**]. This gives a complete disjunction which is founded in the essence of the time sequence, and which we can express thus:

AXIOM VI: If two numbers are different, one of them must be higher than the other.

4. THE ADDITION OF PURE NUMBERS

In order to formulate general propositions about numbers, I shall use the

---

[†] ['die Einordnung in die Zeitfolge'.]

[††] ['Bezeichnung', which Helmholtz here evidently associates with 'Zeichen' ('symbol'); we also translate 'bezeichnen' correspondingly as 'symbolise'. But elsewhere in this article 'Bezeichnung' rather means 'terminology'.]

[*] 'Number theory' investigates number series in which a certain number is always followed again by one, which therefore repeat themselves periodically.

[**] At this stage I still avoid *greater* and *smaller*; this distinction more suitably attaches itself to the concept of cardinal number – of which later.

well-known symbolism of algebra. Each letter of the lower-case Latin alphabet shall be capable of symbolising any arbitrary number, but always the same one within each individual theorem or each individual calculation.

Explanation of symbols. When I have symbolised any number by a letter, e.g. a, I shall symbolise the one immediately following it in the normal series by $(a+1)$.

Thus here this symbol $(a+1)$ shall have, to begin with, no other meaning than the one stated. Parentheses, however, will in general have their usual meaning: that the numbers enclosed within them should first be combined into a number before the remaining prescribed operations are carried out.

The alikeness symbol[†] $a=b$ shall symbolise here, in the pure theory of numbers, only: "a is the same number as b". Therefore, from

$$a=b,$$
$$b=c$$

there follows immediately $a=c$, for the two above equations state that both a and c are the same number b. This establishes, in the pure theory of numbers, the validity of Axiom I for the series of the whole numbers.

5. NUMBERING THE NUMBERS

From now on we consider the normal number series to be established and given. We can now consider its members themselves to be a given series of representations in our consciousness, whose order starting from any arbitrarily chosen member we can again symbolise, using the normal number series beginning with *one*.

DEFINITION. I symbolise as $(a+b)$ that number in the principal series at which I arrive when numbering $(a+1)$ as one, $[(a+1)+1]$ as two, and so on until I have numbered up to b.

The description of this procedure can be summarized in the following

[†] [i.e. what one would normally translate 'the sign of equality'.]

equation[†] (H. Grassmann's axiom of addition):

$$(a+b)+1 = a+(b+1). \tag{1}$$

Elucidation. This equation states that, having started from $(a+1)$ as one and numbered up to b, and thus having found the number symbolised by $(a+b)$, if I continue numbering by one further step, I arrive in the former series at $(b+1)$, and in the second series at the number following $(a+b)$, namely $[(a+b)+1]$. So I thus symbolise $[(a+1)+1]$, mentioned in the definition, also by $[a+(1+1)]$ or $(a+2)$, and further $[(a+2)+1]$ by $(a+3)$, and so on without limit.

In the language of arithmetic we would call this procedure *addition* and the number $(a+b)$ the *sum* of a and b, and a and b themselves the *summands*. However, I draw attention to the fact that in the stated procedure the roles of the magnitudes a and b are not alike, so that proof must first be brought that they can be exchanged without altering the sum. This is to be done further on below. All the same, if we keep this reservation in mind, we can accept this terminology[††] and say that the expression $(a+b)$ prescribes that b should be added to a, and that $(a+b)$ is the sum of a and b, whereby for the moment the order of b behind a must however be kept fixed. We may therefore call a the *first* and b the *second summand*. Correspondingly, each number $(a+1)$ can be termed, by logically applying this terminology, the sum of preceding a and the number one.

The stated procedure of addition must, in the lawlike number series, always yield a result, and indeed always the same one for the same numbers a and b. For each of the steps out of which we have composed the addition $(a+b)$ is an advance by one stage in the series of the positive whole numbers, from $(a+b)$ to $(a+b)+1$, and from b to $(b+1)$. Each single step can be carried out and, according to our presuppositions about the immutable preservation of the number series in our consciousness, each of them must always give the same result. There will therefore certainly exist a number which corresponds to the number $(a+b)$, and only one. This proposition corresponds to the content of axiom IV, when the latter is applied to the pure numbers and to the kind of addition prescribed here.

[†] ['Gleichung'.]
[††] ['Bezeichnung'.]

On the other hand, it follows from the description given of the procedure that $(a+b)$ is necessarily different from a, and indeed higher than a, if b is one of the positive whole numbers.

If c is a higher number than a, I must necessarily reach c by numbering stage by stage upwards from a, and be able to number off what number c is as numbered from a. Let it be the bth such number, then

$$c=(a+b).$$

We wish for later quotation to term this proposition:

AXIOM VII. If a number c is higher than another one a, then I can portray c as the sum of a and a positive whole number b to be found.

THEOREM I. On the sequence of execution of several acts of addition, (*associative law* according to Grassmann).

When I have to add to a sum $(a+b)$ a number c, I obtain the same result as when I add the sum $(b+c)$ to the number a. Or written as an equation:

$$(a+b)+c=a+(b+c). \tag{2}$$

Proof. The proposition, as equation (1) states, holds for $c=1$. It is to be shown that if it is correct for any particular value of c, it is also correct for the immediately following value $(c+1)$.

Now according to Equation (1):

$$[(a+b)+c]+1=(a+b)+(c+1),$$
$$[a+(b+c)]+1=a+[(b+c)+1]=a[b+(c+1)].$$

The latter according to proposition (1).

Thus if proposition (2) holds for the value of c occurring here, the expressions on the left hand side of the first two equations are the same numbers, and consequently also

$$(a+b)+(c+1)=a+[b+(c+1)],$$

which means that the proposition also holds for $(c+1)$.

Then since, as stated before, it holds for $c=1$, it holds also for $c=2$. If it holds for $c=2$, it holds also for $c=3$, and so on without limit[8].

Corollary: Since the two expressions occurring in Equation (2) have the same meaning, we may also omit the parentheses and introduce for both of them the symbolism:

$$a+b+c=(a+b)+c=a+(b+c). \tag{2a}$$

Only we must not alter for the moment the order of a, b, c in these expressions, until we have proved the admissibility of such a permutation.

6. GENERALIZATION OF THE ASSOCIATIVE LAW

Firstly we generalize the symbolism given in (2a).

$$R=a+b+c+d+\text{etc.}+k+l. \tag{2b}$$

shall symbolise a sum in which the individual additions are executed in the order in which they are written, and to abbreviate the symbolism let

$$m+R=m+a+b+c+d+\text{etc.}+k+l,$$

whereas

$$m+(R)=m+(a+b+c+d+\text{etc.}+k+l).$$

On the other hand, according to the sense of this notation:

$$(R)+m=R+m.$$

Other capital Latin letters shall be used in the same sense as R.
Then:

$$R+b+c+S=[(R)+b+c]+S,$$

because they are expressions having the same meaning. On the other hand, by Equation (2a):

$$(R)+b+c=(R)+(b+c).$$

Therefore

$$R+b+c+S=[R+(b+c)]+S=R+(b+c)+S,$$

i.e. instead of adding all the terms in their order, one can first combine two arbitrary intermediate ones into a sum. After this has happened, only one single number stands for the sum $(b+c)$ just formed, and one can go on in the same way to combine any other arbitrary pair of consecutive numbers, and so on.

Thus even with arbitrarily many terms one can alter the order in which the additions prescribed by the individual + signs are carried out, without the total sum altering.

THEOREM II (*commutative law* according to H. Grassmann): If in a sum of two summands one of the summands is one, their order can be reversed. To this there corresponds the equation:

$$1+a=a+1.\tag{3}$$

Proof. The equation is correct for $a=1$. It has to be shown besides, that if it is correct for any particular specified value of a it is also correct for $(a+1)$. Now by Equation (1):

$$(1+a)+1=1+(a+1).$$

By assumption, Equation (3) is to hold for a, consequently

$$(1+a)+1=(a+1)+1.$$

From these two equations it follows that

$$1+(a+1)=(a+1)+1,\tag{3a}$$

which was to be proved.

Since the proposition is correct for $a=1$, it is also correct for $a=2$, and since it is correct for $a=2$ it is also correct for $a=3$, and so on without limit.

THEOREM III. In any sum of two summands the order of the summands can be altered, without altering the number corresponding to the sum. Or written:

$$a+b=b+a.\tag{4}$$

The proposition is correct for $b=1$ by Theorem II. If it is correct for a specific value of b, it is also correct for $(b+1)$. For by Theorem I

$$(a+b)+1=a+(b+1),$$

and by our presupposition

$$(a+b)+1=(b+a)+1=1+(b+a)=(1+b)+a=(b+1)+a.$$

Of the last three steps, the first and last are made by Theorem II, Equation (3) and the intermediate one by Theorem I, Equation (2). Consequently

$$a+(b+1)=(b+1)+a,$$

as was to be proved.

From the proposition

$$a+1=1+a$$

there therefore follows

$$a+2=2+a,$$

from this again

$$a+3=3+a,$$

and so on without end.

Proof of Axiom V. If a and f are different numbers, we can – as shown in Axiom VII – always specify a positive whole number b such that[†]

$$(a+b)=f.$$

Then:

$$c+f=c+(a+b)=(c+a)+b.$$

According to this, $(c+a)$ is then necessarily different from $(c+f)$, that is: different numbers added to the same number give different sums.

Since, however by Theorem III

$$c+f=f+c,$$
$$a+c=c+a,$$

this last conclusion can also be written

$$(f+c)=(a+c)+b,$$

that is: the same number added to different numbers gives different sums.

From this there follows the important proposition for the theory of subtraction and equations: that two numbers which yield the same sum when the same number is added to each of them, must be identical.

[†] [i.e. if it is f which is 'higher' than a.]

7. Permuting the Order of Arbitrarily Many Elements

In what we have described so far, as a method of numbering off for the purpose of addition, two series of numbers, which retained their normal sequence, were pairwise combined with each other, so that $(n+1)$ was coordinated with 1, $(n+2)$ with 2, etc. Only the starting points of the two sequences in the number series were different.

We now wish to consider the more general case of coordinating the elements of two series, of which one is to preserve a certain sequence, and can therefore be portrayed by number symbols, while the other may have variable sequence. As symbols for the latter we wish to use here the letters of the Greek alphabet. These indeed also have a sequence impressed on our memory, as given in the ordinary arrangement of the alphabet. But we want to use this only as one sequence among many other possible ones, which is distinguished by accidental factors and whose habitual recollection facilitates an overall view, and want otherwise to permit ourselves to alter it arbitrarily. On the other hand, we demand that in the alterations to be made in the sequence of these elements, no element shall be omitted and none repeated. We can most easily check this in our memory, if we stipulate that the group should contain as elements all letters preceding a specific letter, e.g. κ, in the received order of the alphabet.

8. Transposing Two Consecutive Elements of a Series

If two elements, e.g. ε and ζ, are coordinated with two consecutive numbers n and $(n+1)$, then n can be linked either with ε or with ζ; this gives the two following manners of coordination:

$$
\begin{array}{cc}
n & (n+1) \\
\varepsilon & \zeta
\end{array}
$$

or

$$
\begin{array}{cc}
n & (n+1) \\
\zeta & \varepsilon
\end{array}
$$

If we substitute for the first of these two orders the other one, and leave unaltered all the other coordinated pairs consisting each of one number and one letter, then no number goes without its coordinated letter and no letter without its coordinated number, we do not repeat and also do not omit any letter. So if the series which contained the first two above-mentioned pairs was a group without gaps or repetitions before the transposition, it is also thus with the series yielded by the transposition.

By suitably repeating such transpositions of neighbouring elements, I can make *any arbitrarily chosen* element of the group the first in the series without producing repetitions or gaps in the series. For if the chosen element ξ is the nth, I can exchange it with the $(n-1)$th, then with the $(n-2)$th, etc., so that its location number becomes lower and lower, until I finally arrive at the lowest position number in the group, namely 1.

In the same way, I can make any element of the series whose position number is higher than m the m-th member of the group without producing gaps or repetitions. In this latter procedure those members of the series whose position numbers are lower than m will retain these numbers unaltered.

From this there ensues the following: by continued exchange of neighbouring members of a group, I can bring about *any possible sequence* of its members without omitting or repeating elements. For I can arbitrarily prescribe which shall become the first member of the series, and achieve this by the given procedure. Then I can prescribe which shall become the second and equally achieve this. The element which has just become first will not thereby be dislodged from its position. Then I can specify the third, etc., up to the last one.

THEOREM IV. Attributes of a series of elements which do not alter when arbitrary neighbouring elements are exchanged in order with each other, are not altered by any possible alteration of the order of the elements.

This leads us in the first place to the *generalization of the commutative law of addition*.

The capital letters may again, as with the generalization of the associative law, denote sums of arbitrarily many numbers. Then, by the generalized associative law, we have:

$$R+a+b+S=R+(a+b)+S.$$

According to the commutative law for two summands

$$a+b=b+a.$$

Therefore, since according to this $(a+b)$ is the same number as $(b+a)$:

$$R+a+b+S=R+(b+a)+S=R+b+a+S.$$

The latter again by the associative law.

Since, according to this, one can exchange with each other two arbitrary consecutive elements without altering the result of the sum, one can by Theorem IV permute all of them with one another, and bring them into any arbitrary order, without altering the sum.

With this one has proved the five basic axioms of addition for what we made the basic concept of addition, and derived them from it. It now remains to show that this concept coincides with the one which issues from ascertaining the cardinal number [†] of numerable objects.

This leads us first of all to the concept of the cardinal number [9] of the elements of a group. If I need the complete number series from 1 to n in order to coordinate a number with each element of the group, then I call n the *cardinal number* of the members of the group. The discussion preceding the formulation of Theorem IV shows that the cardinal number of the members remains *unaltered by changes in their order*, when omissions and repetitions of them are avoided.

This theorem is now applicable to real objects for which α, β, γ, etc. may be regarded as temporarily given names. Only, these objects must – at least for as long as the result of an executed numbering is to be valid – in fact fulfil certain conditions in order to be numerable. They should not disappear, or merge with others, none of them should split into two, no new one should be added, so that each name given in the form of a Greek letter will also continue to correspond to one and only one circumscribed object, which remains recognizable as a single one[*]. Whether these conditions are obeyed for a specific class of objects can naturally only be determined by experience [10].

From the method of numbering described above it follows immedi-

---

[†] ['cardinal number' translates 'Anzahl'; in some contexts 'totality' would give a more natural rendering.]

[*] With this paper already in the press, I learn that Prof. L. Kronecker, in a lecture during the past winter semester, developed in a similar way the concepts of number and cardinal number.

ately, by the concept of addition already laid down, that the *total number* of the members of two groups which have no member in common equals [†] the *sum of the cardinal numbers* of the members of the two single groups. In order to find the total number, one would be able to number firstly through the one group. If it has p members, the number $(p+1)$ would fall upon the first member of the other group, $(p+2)$ upon the second one, and so on. Thus the total number of members in the two groups will be found by exactly the same numbering procedure as will the sum – as defined above – of the two numbers which give each the cardinal number of elements in one group.

The concept of addition described above therefore indeed coincides with the concept of it which proceeds from determining the total cardinal number of several groups of numerable objects, but has the advantage of being obtainable without reference to external experience.

One has hereby proved, for the concepts of number and of a sum – taken only from inner intuition – from which we started out, the series of axioms of addition which are necessary for the foundation of arithmetic; and also proved, at the same time, that the outcome of this kind of addition coincides with the kind which can be derived from the numbering of external numerable objects.

Starting from here, the theory of subtraction and multiplication is developed without further difficulties, by defining the *difference* $(a-b)$ as that number which must be added to b in order to obtain a as sum, and multiplication as the addition of a cardinal number of like numbers. I need only refer here to Grassmann, who defines the multiplication of pure numbers by the two equations:

$$1 \cdot a = a,$$
$$(b+1) \cdot a = b \cdot a + a.$$

With regard to subtraction, it should only be noted that one can also continue the numbers, as symbols of a sequence, without limit in a descending direction, by going backwards from 1 to 0, thence to (-1), (-2), etc. and treating these new symbols just like the positive whole numbers which we previously used alone. Then the difference between two numbers always has one and indeed only one meaning; it is therefore determined unambiguously.

[†] ['gleich ist'.]

The *agreement* as well as the *difference* between the laws of addition and multiplication must also be discussed here, because of what follows.

The associative law and the commutative law hold for both operations. As we have seen:

$$(a+b)+c=a+(b+c)$$
$$a+b=b+a.$$

But also likewise:

$$(a\cdot b)\cdot c=a\cdot(b\cdot c)$$
$$a\cdot b=b\cdot a.$$

A difference between the basic properties of the two operations does not emerge until one combines, by each of them, a cardinal number n of like numbers a. Combined by addition, these give the product $n\cdot a$, which is itself again subject to the commutative law:

$$n\cdot a=a\cdot n.$$

By multiplication of n like factors, on the other hand, one gets the power a^n, in which – except in special cases – a and n cannot be exchanged without altering the value of the power.

An analogy equally emerges in one case when we combine each of these operations with the next higher one:

$$c\cdot(a+b)=(c\cdot a)+(c\cdot b)^{11}$$
$$c^{a+b}=c^a\cdot c^b.$$

But the analogy is lost under commutation, as the same relation no longer holds for $(a+b)^c$ on the one hand and $a^c\cdot b^c$ on the other[†].

9. DENOMINATE [CONCRETE] NUMBERS[††]

If we need to number unlike objects in the manner discussed above, it is generally only in order to check their full number. Of much greater importance and wider application is the numbering of like objects. Such objects, which in some or another specific respect are alike and get

---

[†] [i.e. these two expressions usually denote different numbers, whereas $(a+b)\cdot c=(a\cdot c)+(b\cdot c)$ always. (Helmholtz' point here is obscured in the original by his misleadingly writing the latter equation in the text instead of $c\cdot(a+b)=(c\cdot a)+(c\cdot b)$ as above)].

[††] ['benannt', 'Benennung.' In English one more commonly speaks of a "concrete number"].

numbered, we call the *units* of the numeration, the cardinal number of them we term a *denominate*[†] *number*, and the particular kind of units with it gathers together the *denomination*[†] of the number.

A cardinal number, as we have seen above, is resolvable into parts which are *additively* gathered together in the whole. The sum of two denominated numbers of like denomination is the total number of all their units, thus necessarily again a denominate number of the same denomination. When we have to liken two different groups of different cardinal numbers, we term the one having the higher cardinal number the *greater*, and that having a lower cardinal number the *smaller*. If both have the same cardinal number, we term them *alike*.

Objects, or attributes of objects, which when likened with similar ones allow the distinction into greater, alike or smaller, we call *magnitudes*. If we can express them by a denominate number, we call the latter the *value* of the magnitude, and the procedure by which we find the denominate number *measurement* of the magnitude. We incidentally, in many investigations carried out in practice, only succeed in reducing the measurement to units which are arbitrarily chosen or are given by the instrument used; the numbers which we find have then only the value of *proportional numbers*[††], until those units are reduced to universally known ones (*absolute units of physics*). However, these universally known units are not to be defined by their concept, but can only be displayed in particular natural bodies (weights, measuring rods) or particular natural processes (day, pendulum beat)[12]. The fact that they are more universally known by tradition among men does not alter the business and the concept of measuring, and appears in contrast with this as only an incidental feature.

In the following we shall have to investigate in which circumstances we can express magnitudes by denominate numbers, i.e. find their value, and what we attain thereby as regards factual knowledge. For that, however, we must first discuss the concepts of alikeness and of magnitude in respect of their objective meaning.

10. PHYSICAL ALIKENESS[†††]

The special relationship which may exist between the attributes of two

---

[†] ['benannt' and 'Benennung'.]   [††] ['Verhältniszahlen', i.e. ratios.]   [†††] ['Gleichheit', normally 'Equality'.]

objects, and to which we give the name 'alikeness', is characterized by Axiom I as already adduced above: if two magnitudes are both alike with a third, they are alike amongst themselves. This implies at the same time that the relationship of alikeness is a mutual one. Because from

$$a = c,$$
$$b = c$$

it follows just as well that $a = b$ as that $b = a$.

Alikeness between the likenable attributes of two objects is an exceptional case, and can therefore be indicated by factual observation only in that the two like objects, when meeting or interacting in suitable conditions, allow the observation of a particular outcome which does not as a rule occur between other pairs of similar objects. We wish to term the procedure by which we place the two objects in suitable conditions for observing the said outcome, and ascertaining whether or not it takes place, the *method of likening*[†].

If this procedure of likening is to give sure information on alikeness or difference for a specific attribute of the two objects, its outcome must exclusively and solely depend upon the condition that both objects possess the relevant attribute in the specific measure, always presupposing that the likening procedure is properly applied.

It follows, from the axiom adduced above, firstly that the outcome of this likening must *remain unaltered* if the two objects are interchanged. It moreover follows that if the two objects a and b prove to be alike, and it has been found by previous observation using the same method of likening that a is also alike with a third object c, then the corresponding likening of b and c must also show these to be alike.

These are requirements which we have to lay down for the relevant method of likening. *Only* those kinds of procedure which fulfil the said requirements are capable of demonstrating alikeness[13].

From these presuppositions, it follows that "like magnitudes can be substituted for each other" in the first place for the outcome on whose observation we rely for ascertaining their alikeness. However, the alikeness of further effects or relationships of the relevant objects may also be connected, by natural laws, with alikeness in the case so far discussed, so that the relevant objects may be interchangeable also in these other

[†] ['Vergleichung' , ordinarily translated 'comparison' .]

respects. We are accustomed to expressing this linguistically as follows: we objectivise, as an attribute of the objects, the capacity they have for bringing about the outcome decisive for the first kind of likening; then we ascribe like magnitude of that attribute to the objects found to be alike, and characterise the further effects in which alikeness is preserved as effects of that attribute, or as empirically dependent upon that attribute alone. The sole meaning of such an assertion is always this: that objects which have proved to be alike for the kind of likening which decides concerning alikeness of this particular attribute are also mutually substitutable, without altering the outcome, in the further cases mentioned.

Magnitudes whose alikeness or nonalikeness is to be decided by the same method of likening are termed by us 'alike in kind' [†]. In our separating, by abstraction, the attribute whose alikeness or nonalikeness is hereby determined from everything whereby the objects otherwise differ, there remains for corresponding attributes of different objects only distinction by magnitude [*].

Let me elucidate the sense of these abstract propositions by some well-known examples.

Weights. When I place two arbitrary bodies on the pans of a true balance, the balance will generally not be in equilibrium [††], but one pan will sink.

Exceptionally, I shall find certain pairs of bodies a and b which, when placed on the balance, will not disturb its equilibrium. If I then exchange a and b, the balance must remain in equilibrium. This is the well-known test of whether the balance is true, i.e. whether equilibrium of this balance indicates alikeness of the weights.

Finally it is confirmed that if the weight of a is alike not merely with the weight of b but also with that of c, then also $b = c$. The equilibrium of weights on a true balance, therefore, is indeed the basis of a method for determining a kind of alikeness.

[†] ['gleichartig', i.e. homogeneous.]

[*] H. Grassmann's definition of alikeness: "Those things are alike about which the same can always be asserted or, more generally, which can be mutually substituted in every judgement" would demand that in every single case where alikeness is inferred this most general requirement, which is exposed to misinterpretation, should be applied [14].

[††] ['im Gleichgewichte', suggesting the meaning 'weighted alike'.]

The bodies whose weight we liken may incidentally consist of the most varied materials, and have the most varied forms and volumes. The weight which we make alike is only an attribute of them which has been separated by abstraction. If we call the bodies themselves weights, and call these weights magnitudes, then we should do this only when we can disregard all their other properties. This has its practical sense whenever we observe or bring about processes in the course of which only this one attribute of the bodies involved is under consideration, i.e. in which bodies which yield an equilibrium on the balance are mutually substitutable. This is for instance the case when we measure the inertia of the bodies concerned. That bodies of like weight are also of like inertia, and are substitutable for each other also in the latter connexion, does not however follow from the concept of alikeness, but only from our knowledge of this particular law of nature for this particular connexion[15].

[11. DISTANCES BETWEEN TWO POINTS][†]

The simplest geometrical structure for which a magnitude is specifiable is the distance between a pair of points. But to give a specific value to the distance for at least the time of the likening which measures it, the points must be fixedly linked as e.g. the tips of a pair of dividers. The well-known method of likening distance for two pairs of points consists in our investigating whether or not they can be brought into congruent coincidence. Experience confirms that this method is suitable for ascertaining alikeness, that the congruence always recurs in any situation and on exchanging the two point pairs in any arbitrary manner, that two point pairs which are congruent with a third are also congruent amongst themselves. In this way we can form the concept of like distances or separations[16].

Thence one can proceed to the concept of the straight line and its length. Think of two fixed points through which there is to pass a line. A *straight* line is one in which no point can alter its situation without altering at least one of its distances from the fixed points. A *curved* line, on the other hand, can be rotated about two of its points, whereby the

---

[†] [In the original the remainder of this section is erroneously made a separate section; in fact "distances between two points" is merely one of the "well-known examples" promised above by Helmholtz.]

situation of the remaining points is indeed altered, but not their distance from the two fixed points. We make the *length* of two bounded straight lines alike if the distance between the end points is alike, thus when the latter can be placed in congruence, whereby the lines too coincide congruently. To this extent the concept of length gives something more than does the concept of distance. If we think of two point pairs *a,b* and *a,c*, of differing distance, which coincide at *a* and which are placed in a straight line, so that a portion of this line is common to both, then either *b* falls upon the line *ac* or *c* upon the line *ab*. This gives an opposition corresponding to that of greater and smaller, whereas the concept of distance gives immediately only that of like and unalike.

Time measurement presupposes the finding of physical processes which, repeating themselves in like manner exactly and under like conditions, if they have begun at the same moment also end simultaneously[†]. For example: days, pendulum beats, the running out of sand and water clocks. Here the justification for assuming unaltered duration of repetition lies only in the circumstance that all the various methods of time measurement, if carefully executed, always yield concurring results. If two such processes *a* and *b* begin and end simultaneously, they take place not only in like time but in the same time. In respect of time there is no possible distinction between the two of them, and therefore also no possible exchange. And if a third process *c*, beginning simultaneously with *a*, ends in the same time, then it also does so with the simultaneously proceeding process *b*.

We liken the *brightness* of two visible areas by bringing them up to each other – so that any demarcation between them disappears apart from difference in brightness – and then checking whether there still remains a recognizable boundary between them[17].

We liken the *pitch* of tones, if it is a matter of small differences, by using the phenomenon of beats, which must be absent if the pitches are alike. We liken the *intensities of electric currents* with the differential galvanometer, which remains unaffected by them if they are alike[18]. And so on.

Thus for the task of ascertaining alikeness in various respects, the most varied physical means have to be sought out. But all of them must fulfil the requirements laid down above, if they are to prove an alikeness.

[†] ['gleichzeitig' , with the literal meaning: 'alike regarding time' .]

The first axiom – "If two magnitudes are both alike with a third, they are alike amongst themselves" – is thus not a law having objective significance; it only determines which physical relations we are allowed to recognize as alikeness.

To quote some instances in which this axiom concerning the alikeness of the third object even underlies something executed mechanically, I may mention the grinding of plane glass surfaces. If two of these are ground down by rotating one of them continuously against the other, both may become spherical – the one concave and the other convex. If three are rotated against one another alternately, they must finally become plane. In the same manner, the edges of exact metal rulers are made straight by grinding them down against one another in threes.

12. ON ADDITIVE PHYSICAL CONNEXION* OF MAGNITUDES ALIKE IN KIND

Likening magnitudes, as discussed by us so far, answers the question of whether they are like or unlike; but should they be different, it does not yet give any measure for the magnitude of their difference. However, if the magnitudes concerned are to be completely specifiable by denominate numbers, the greater of the two numbers must be portrayable as the sum of the smaller and their difference. We have to investigate, therefore, under what conditions we may express a physical connexion between magnitudes alike in kind as an addition.

The manner of connexion involved here will in general depend upon the kind of magnitudes to be connected. We add e.g. weights simply by putting them in the same pan of a balance. We add periods of time by letting the second one begin at the exact moment when the first one ends; we add lengths by placing them one next to the other in a certain manner, namely in a straight line. And so on.

1. *Alikeness in kind of the sum and the summands.* As this connexion is to involve magnitudes *of a specified kind*, its result should not alter if I exchange one or more of the magnitudes with magnitudes alike in magnitude and in kind. For they will then only be replaced with the

---

* 'Connexion' ['Verknüpfung'] is a Grassmannian term, though his use of it is predominantly subjective, ours objective alone.

same number having the same denomination as they already had themselves. But the result of connexion, if it is to count as the sum of the connected magnitudes, must also be alike in kind with the parts, since the sum of two denominate numbers is again, as already mentioned above, a number of the same denomination. Therefore, the issue of whether the result of connexion stays alike[†] when parts are exchanged, must be decided *by the same method of likening* with which we ascertained the alikeness of the parts to be exchanged. This is the factual significance of the requirement that the sum of magnitudes alike in kind must be alike in kind with its summands.

Thus e.g. in a sum of weights I can replace the individual pieces with ones of another material but like weight. Then the sum retains like weight, but its remaining physical attributes can alter.

2. *Commutative law.* The result of addition is independent of the order in which the summands are connected. The same must hold for physical connexions which are to be regarded as additions.

3. *Associative law.* The conjoining of two magnitudes alike in kind can also take place physically by replacing them both with one undivided magnitude of the same kind and alike [in magnitude] with their sum. In this way, the two are then additively united before all the others. Thus e.g. in weighing, a five-gram piece can be substituted for five one-gram pieces. The result of connexion should therefore not alter if I introduce, instead of some of the magnitudes to be connected, others which are alike with the sum of these.

It can incidentally be shown that if the first two requirements are universally fulfilled, the third one is fulfilled too.

Think of the elements ordered successively in the sequence in which, as a first case, they are to be connected with one another by affixing each one to the result of connecting those preceding it, as we prescribed above for the addition of $(a + b + c + \text{etc.})$. If now, as a second case, any of these magnitudes are required to be connected before the others, then by the commutative law – which is to hold by assumption – we can put them in first and second place, where they are then to be connected before the

---

[†] [i.e. remains the same – 'Gleichbleiben des Ergebnisses' .]

others, without our altering the result. Thereupon, according to our first condition above, we can also replace the result of this connexion with another undivided object which, considered as a magnitude alike in kind, is alike in magnitude.

After that, we can take the next two magnitudes – or sums of magnitudes – to be connected, and bring them in their turn into the first two places, and so on, until all are connected in the prescribed sequence. By none of these operations do we alter the magnitude of the final result of the connexions.

A *physical method* of connecting magnitudes alike in kind can be *regarded as addition*, if the result of the connexion – when likened as a magnitude of the same kind – is not altered *either* by exchanging individual elements with each other, *or* by exchanging terms of the connexion with like magnitudes of like kind.

With our having found the method of connecting the magnitudes concerned, in the way just described, it now also follows which are greater and which smaller. The product of additive connexion, the whole, is greater than the parts of which it is composed. Regarding those simplest magnitudes which we have dealt with from earliest youth, such as times, lengths and weights, we have never doubted about what was greater and what smaller, because we have indeed known additive methods of connecting them all along. It should however be borne in mind here that the method of likening, as described above, generally tells us only whether the magnitudes are like or unalike.

If two magnitudes x are alike, then all functions of them formed by calculation in like manner are also alike. Of these some will increase and others decrease as x increases. Which of these functions permit an additive physical connexion is only to be decided by particular experience.

Those cases in which two kinds of additive connexion are possible are then instructive. Thus, by exactly the same method of likening, we determine both whether two wires are of like electrical resistance w and whether they are of like conductance λ, because

$$w = \frac{1}{\lambda}.$$

We add resistance by joining the wires in succession so that the conducted electricity must flow through them all one after another. We add the conductance of the wires by placing them side by side and joining them all up together at the one end and at the other. Thus we objectivise here, as physical magnitudes, two different functions of the same variable. If a wire has greater resistance, then it has lower conductance, and conversely. Thus the question of what is greater and what smaller will be answered for the two of them conversely. In the same manner electric condensers (Leyden jars) can be joined side by side or in succession. In the former case the capacity is added for like charge, in the latter one the voltage (potential). The former grows when the latter decreases.

All the same, we should not be surprised if the axioms of addition are verified in the course of nature, since we recognize as addition only those physical connexions which satisfy the axioms of addition.

13. THE DIVISIBILITY OF MAGNITUDES AND UNITS

So far we have not yet needed to analyse magnitudes into units. The concept or magnitude, as of alikeness in magnitude and of addition of magnitudes, was obtainable without such an analysis. However, we in fact attain the greatest simplicity in portraying magnitudes only on resolving them into units and expressing them as denominate numbers.

Magnitudes which can be added are in general also divisible. If every occurring magnitude can be regarded as additively composed, by the addition procedure valid for magnitudes of this kind, out of a cardinal number of like parts, then by the associative law of addition each of these magnitudes can be replaced, wherever only its value is of account, with the sum of its parts. It is in this way then replaced with a denominate number, and other magnitudes of like kind with other cardinal numbers of the same parts. The description of the individual magnitudes of like kind can then be conveyed, to a listener acquainted with the like parts chosen as units, by simply enunciating the numbers.

If the occurring magnitudes are not expressible without remainder using the chosen units, then one divides the units again in the known manner, and can in this way give a specification of the value of any of the occurring magnitudes to any arbitrary precision. Complete precision is of course only attainable for rational proportions.

Irrational proportions may occur in real objects [19]; yet one cannot ever portray them with complete accuracy in numbers, but only enclose their values between arbitrarily reducible limits. This narrowing down[20] between limits suffices for computation of all functions such that if the changes in the magnitudes upon which they depend become smaller and smaller, then the values of the functions themselves undergo smaller and smaller changes, which in the end can fall below any statable finite amount. This holds in particular for the calculation of differentiable functions of irrational magnitudes. On the other hand, one can of course also form discontinuous functions for calculating which it does not suffice to know the limits between which the irrational value lies, however narrowly these may be drawn. In respect of those functions, the portrayal of irrational magnitudes using our number system always remains insufficient. But in geometry and physics we have not yet encountered such kinds of discontinuity.

14. QUANTITATIVE DETERMINATION OF PROPERTIES (PHYSICAL CONSTANTS, COEFFICIENTS)[†]

Apart from the magnitudes discussed so far, which are directly recognizable as such because they can be conjoined by addition, there still remains a series of other relationships, also expressible using denominate or non-denominate numbers, for which an additive conjunction with ones alike in kind is not yet known. They are found whenever there occurs a relation, obeying natural laws, between additive magnitudes in processes which are influenced by the peculiarities of some particular substance or particular body, or by the particular way in which the process concerned was initiated.

Thus e.g. the law of refraction of light declares that a particular relationship exists between the sines of the angles of incidence and refraction of a light ray of a particular wavelength, when this enters a given transparent substance from a vacuum. But the number expressing this relationship is different for different transparent substances, and thus denotes a property of it – its refractive index. Specific gravity, thermal conductivity, electrical conductivity, thermal capacity and so on are similar magnitudes. To these one may add those quantities (integration constants of dynamics) which remain unaltered during the undisturbed

---

[†] [In this section, 'quantity' translates 'Wert' (normally 'value'). In the original this section is misleadingly made a subsection of the preceding one.]

course of the motion, once it has been initiated, of a limited system of bodies[21].

Physics has gradually succeeded in reducing all of these quantities to units which can be composed – by multiplication, raising to a power and the inverses of these operations – out of the three fundamental units of measurement for time, length and mass[22].

The distinction between these quantities and additive magnitudes is not strictly adhered to in the language of physicists and mathematicians. The former too are often called magnitudes, as one expresses them using denominate numbers, though the term *coefficient* characterises their physical nature relatively better. But this is not an essential distinction, for occasionally new discoveries can lead to ways of additively conjoining such coefficients, whereby they would move into the class of directly determinable magnitudes. To some extent this distinction indeed corresponds to the one which metaphysicians of an earlier period wished to state using the antithesis of *extensive* and *intensive* magnitudes. P. du Bois-Reymond calls the former *linear* and the latter *non-linear* magnitudes[23].

On the other hand, however, it emerges from the given derivation that one must have first formed additive magnitudes before one can determine coefficients. For the equation expressing a natural law cannot quantitatively determine a coefficient, until all other magnitudes occurring in it are already determined as magnitudes. The determination of additive magnitudes must therefore always come first, before one can find the values of non-additive ones.

15. THE ADDITION OF MAGNITUDES NOT ALIKE IN KIND[†]

A major role is played in physics by those objects which, when compared by different methods of likening, manifest simultaneously two, three or even several magnitudes of different kinds, all of which are added when the objects are combined in the same physical way. Amongst these belong firstly the very large number of magnitudes oriented in space which occur in physics, i.e. magnitudes having a specific value and at the same time a specific orientation, yet ones which one can represent as a combination

---

[†] [i.e. nonhomogeneous magnitudes.]

of several components (two in a plane, three in space) having fixed orientations. The overall view of the situation is made most simple, in general, by choosing as parallel to three rectangular coordinate axes those components which are to be connected, in the way prescribed by the law of the parallelogram of forces, to form a resultant[24]. To this class there belong displacements of a point in space, its velocity and acceleration; corresponding to the latter the force propelling the point; moreover angular velocities[25] and vanishingly small rotations, the current velocities of heavy fluids, electricity and heat, magnetic moments, and so on.

To conjoin additively one sums the components having like orientation, then these sums can be brought together to form a resultant. All physical connexions of such magnitudes for which the outcome depends only upon the magnitude and orientation of the final resultant, can be regarded as based upon additive connexions of this kind. This was done for two dimensions by Gauss in his geometrical interpretation of imaginary magnitudes[26], for several dimensions by H. Grassmann[27] as the addition of geometrical extents†, and by R. Hamilton in the theory of quaternions[28]. The commutative law must be satisfied here too. Thus we can combine, into a resultant rotation, infinitely small rotations of a fixed body about two different axes, and combine angular velocities as well, but not finite rotations – because with these it is no longer a matter of indifference whether one rotates first about axis a and then about axis b, or conversely.

Yet even in mixing coloured light a similar relationship occurs. Any quantity of coloured light can be portrayed, in respect of the corresponding sense-impression, as a composition of three quantities of light of suitably chosen basic colours. Mixing several colours then has the same effect upon the eye as would a composition of three quantities of light, from the three basic colours, obtained for each particular basic colour by adding the corresponding quantities occurring in the composition of the various particular colours. This forms the basis for the possibility of portraying the laws of colour mixing geometrically with centre-of-gravity constructions, as first proposed by Newton[29].

---

† ['geometrische Strecken'; the addition of vectors is meant.]

16. THE MULTIPLICATION OF DENOMINATE NUMBERS

A denominate number $(a \cdot x)$, where x is to stand for the kind of unit and a for the cardinal number of such units, can be multiplied by a pure number n. This simply falls under the definition, adduced above, of the product as the sum of n like summands a. Since the sum of summands alike in kind is a magnitude alike in kind with them, so also is the product $(n \cdot a)$ a magnitude of the same denomination as a. The commutative law applies to this product, inasmuch as

$$n \cdot (a \cdot x) = a \cdot (n \cdot x),$$

i.e. one can also regard a as a pure number, and form new denominate summands $(n \cdot x)$.

In the same way, the law of multiplication of a sum is immediately given:

$$(m + n) \cdot (a \cdot x) = m \cdot (a \cdot x) + n \cdot (a \cdot x)$$
$$n \cdot (a \cdot x + b \cdot x) = n \cdot (a \cdot x) + n \cdot (b \cdot x).$$

Thus the multiplication of denominate numbers by pure numbers remains wholly within the framework of the definitions and propositions which were derived above for the multiplication of pure numbers by one another.

It is otherwise with the multiplication of two or more denominate numbers. This has a sense only in certain cases, when there can exist, between the units concerned, special physical connexions subject to the three laws of multiplication:

$$a \cdot b = b \cdot a$$
$$a \cdot (b \cdot c) = (a \cdot b) \cdot c$$
$$a \cdot (b + c) = (a \cdot b) + (a \cdot c).$$

In geometry, the best known examples of multiplicative combinations of this kind are the values of the area of a parallelogram and the volume of a parallelopiped, expressed by the product of two or three lengths, namely of a side and one or two altitudes. But physics forms a great number of such products of different units, and has correspondingly also examples of quotients, powers and roots of them. If we denote a length by l, a time by t and a mass by m, then some of these with their denominations[30] are:

area	l^2
volume	l^3
velocity	l/t
motive force	$m \cdot l/t^2$
work	$m \cdot l^2/t^2$
pressure on a surface	$m/l \cdot t^2$
tension in a surface	m/t^2
density	m/l^3
quantity of magnetism	$(l/t) \cdot \sqrt{m \cdot l}$
magnetic force	$(l/t) \cdot \sqrt{m/l}$

Most of these combinations are based upon the determination of co-efficients, but many of these magnitudes can indeed also give us additive physical connexions, such as velocities, currents, forces, pressures, densities and so on. However, none of these multiplicatively defined units are alike in kind with those from which they are created, and they acquire a sense only through knowledge of particular geometrical or physical laws.

One should mention here the particular variety of multiplication which H. Grassmann laid down for oriented magnitudes in his theory of extension, and which is also made fundamental in the theory of quaternions[31]. This lays down a different commutative law, namely

$$a \cdot b = -b \cdot a$$

and indeed gives such a great simplification in the symbolisation, although not in the calculation of the values produced by the interaction of differently oriented magnitudes.

In this kind of calculation, the product of two extents [†] is the area of the parallelogram having the two of them as sides; however, one considers the surface bounded by the parallelogram to be positive on one side and negative on the other. When considering the one side of the surface, I must make a right hand rotation in going from side a to side b through the angle between them; when considering the other side, I conversely make a left hand rotation from b to a. This is the basis of the difference in the sequence $(b \cdot a)$ and $(a \cdot b)$[32].

---

† [i.e. the vector product.]

It suffices to have mentioned here these forms of calculation and characterised their relation to the forms of calculation in pure number theory, since the present study has only been set the task of showing the significance and justification of calculation with pure numbers, and the possibility of applying them to physical magnitudes.

Our portraying some physical relationship as a magnitude, consequently, can only ever be based upon empirical knowledge of certain aspects of its physical behaviour in meeting and interacting with others. Thus I also conceive of the congruence of two spatial magnitudes which occur in bodies or are bounded by bodies, in the sense of my previous studies on the axioms of geometry, as a physical relationship to be ascertained empirically. We must know *firstly* the method of likening the magnitudes concerned, whereby they are characterised in kind, and *secondly* either the methods of additive connexion or the natural laws in which they occur as coefficients, if we are to be able to express them using denominate numbers.

The great simplification, and comprehensive clarity, of understanding which we achieve by reducing to quantitative relationships the variegated multiplicity of things and changes available to us, has a deep foundation in the nature of how we form concepts. When we form the concept of a class, we bring together in that concept everything that is alike in the objects belonging to the class. And also when we understand a physical relationship as a denominate number, we have expelled from the concept of its units everything of the differences which adhere to these in actuality. They are objects which we now only consider as instances of their class, and whose effectiveness in the respect under investigation also only depends upon their being such instances. In the magnitudes formed from them, there then remains only the most incidental of distinctions – that of cardinal number.

NOTES AND COMMENTS

[1] From this passage it firstly emerges that Helmholtz' theory of arithmetic is intended to be empiricism. Further on below (p. 75), he calls arithmetic a "method constructed upon psychological facts". Thus he regards not merely outer, but also inner experience as a source of axioms. But it is shown by the characterization given here of his theory as empiricism, and also by his comparing the present paper with his writings on geometry, that in his view arithmetical propositions possess a significance independent of our own subjective characteristics.

One cannot assent to the view, also advocated by Kant (*Prolegomena*, §10)[†], that the axioms of arithmetic belong to the form of intuition of time. The grounds for this opinion would be the one-dimensionality of both the number series and time; and furthermore the circumstance that the axioms of arithmetic can also be applied to objects (experiences) having no relation to anything spatial: for since everything given to us in sensation is given to us in a temporal manner, it seems plausible to relate the axioms of arithmetic to the form of intuition of time. But even simultaneously existing things given in a spatial manner can be numbered. Indeed, one should also take account of the view that objects which are neither spatial nor temporal can be numbered, e.g. laws of nature, concepts, etc. Thus one must at any rate conclude that the axioms of arithmetic possess a higher generality than do those of geometry.

[2] The view that universally valid propositions must find their explanation in psychology, or that their validity is of psychological origin alone, is termed psychologism. Whoever does not see the ground for the validity of such laws in the order of objects (taken in the most general sense), must look for it on the side of the subject – either in the form of cognition in general (apriorism), or in the particular characteristics of the faculty of understanding (psychologism). But inasmuch as psychologism truly wishes to retain the universal validity of the propositions concerned, it only reduces one riddle to another, and is incapable of explaining how those laws prove true for what actually happens (e.g. the validity of physical geometry for what actually exists in space and time).

Helmholtz himself, in the two preceding papers, expounded a decidedly anti-psychologistic view in reference to physical geometry. It might at first seem necessary to abandon this viewpoint as regards arithmetic, for arithmetic and geometry differ in the following respects essential to our present issue:

(1) The objects of geometry are external things, or at any rate representational images which have certain relations to external things. On the other hand, some numerable things are only realized psychically. One can e.g. just as well number experiences having a purely affective nature as one can representational images.

(2) For the fundamental relation with which arithmetic deals, namely coordination[††], reducibility to physical relations is not as illuminating as is the reducibility of the concept of like distances to the concept of coincidence.

Nevertheless, on the grounds adduced above one will not be able to stop at a psychologistic theory of arithmetic. The psychic is rather, as regards the axioms of arithmetic, completely coordinated with the physical, and is involved only as an *object* of our cognition. We shall not be allowed to claim that the nature of our cognition is a ground for the validity of those axioms. This seems also to be the view of Helmholtz.

[3] These empirical properties are stated by Helmholtz on p. 86; see also Schröder, *op. cit.*, p. 15.

[4] If Helmholtz' theory of arithmetic is intended to be empiricism in the same sense as his theory of geometry, then this passage can only mean: in our reflection upon experiences we gain those concepts which possess a significance projecting beyond the psychic, and which can therefore *inter. alia* also be applied to the physical.

[†] [According to Kant, time is the 'form of inner sense' or of 'inner intuition', as space is that of 'outer sense/intuition'; see note I.23 and Translator's Note.]
[††] ['Zuordnung'; in English what P.H. has to say about coordination is usually formulated in terms of 'correspondence' and 'association'.]

[5] One does not need to think of any temporal succession, and can acquire the number series from a coordination which is stipulated once and for all. That which is coordinated to an element is called by us the successor of this element (R. Dedekind, *Was sind und was sollen die Zahlen?* ['What are numbers and what do they mean?'] 3rd ed., Brunswick, 1911, p. 21; see also notes 8 and 9).

[6] Through this explanation we attain understanding of the concept of next number symbol[†]. This does not yet give a definition for the concept of a number in general, as one already realizes from the necessity of using the words 'and so on' (see note 9).

[7] The simplest portrayal of a number consists in production of the corresponding cardinal number of objects, say of a series of points. The decimal number system does not bring about a fundamental improvement, but only an alleviation of a quantitative kind. With very high numbers the decimal portrayal in practice fails. One then employs portrayal by powers of ten. In place of this procedure too one must eventually introduce a new one, with use of another manner of symbolization, and so on *ad infinitum*. Each manner of symbolization becomes unusable at some point, and none is better in principle than the simple portrayal with points.

[8] Here use is made of the principle of 'complete induction'[††], which however is by no means without need of proof. Such a proof was e.g. given by Dedekind, in his essay *Was sind und was sollen die Zahlen?* [*op. cit.*], for the set of natural numbers which was singled out by him, by means of a definition, from an arbitrary infinite set. E. Zermelo gives a definition of finite set (see note 9) and can then prove the induction theorem for the set of all finite sets (numbers).

[9] One sees that Helmholtz reduces the concept of a cardinal number[†††] to the concept of an ordinal number. His account does not contain a definition of finite set, in the sense of a criterion which would allow one to decide whether something was a finite set or not. Let us recall how the concept of a number symbol was introduced (p. 87), and pay special attention to the fact that in the explanation concerned the words "and so on" occurred. We shall thus be able to say that a set is finite only if the set less than it by 1 is finite; but this itself is finite if the set less by 1 more is finite. One would thus have to see whether one arrives at 1 by continued subtraction, i.e. one must already possess the concept of finite set. It seems we must accordingly presuppose the concept 'finite set' as a basic concept[††††].
    Starting from the conceptual constructions of set theory, E. Zermelo has laid down a

---

[†] [i.e. the concept of one symbol's being the successor of another symbol.]
[††] [More accurately, Helmholtz is using the equivalent principle of 'mathematical induction' (in today's terminology).]
[†††] ['Kardinalzahl'; P.H. uses this term synonymously with 'Anzahl'.]
[††††] [The use of 'and so on' can be eliminated, if one defines the series of numbers 'inductivily' as follows: (a) 1 is a number, (b) if n is a number, then n' is the immediately following number.
    The substantial objection against Helmholtz is rather that he takes *numbers* to be simply *numerals* (number symbols), whereas numbers should be some kind of *objects* which numerals *denote*. The Frege-Russell definitions of number, which P.H. discusses in this note, are attempts to state what sort of object numbers are.]

simple definition for finite sets (*Atti del IV Congresso dei Matematici*, Rome, 1908, vol. 2, p. 8). We wish to indicate in an example what is involved.

Let us think of a finite set of things:

$$a\,b\,c\,d\,e\,f\,g.$$

These we can order, thus say in the way we have written them out above: b follows upon a, c upon b, d upon c, and so on. Or in general: in the scheme

$$a\,b\,c\,d\,e\,f$$
$$b\,c\,d\,e\,f\,g$$

the number lying below follows upon the one lying above, so that one obtains the pairs

$$ab,\ bc,\ cd,\ \text{etc.}$$

We now wish to split our set, in some way or another, into two parts, e.g.:

$$a\,g\,f / b\,c\,d\,e.$$

Then I can always find, from the pairs above, at least one pair which is divided between the two subsets, e.g. in our present case: ab, ef. For if this were not possible, then one subset must of course contain ab, thus also bc, thus also cd, etc. – thus the whole set, and nothing would remain for the other subset.

Now if the set can be so ordered that in every splitting at least one pair is separated, then the set is finite according to Zermelo. This definition proves itself to be very fruitful, inasmuch as the most important propositions of number theory can be obtained from it straight away (see note 8).

The significance of the concept of coordination in providing a foundation for number theory was recognized first by Dedekind in his essay *Was sind und was sollen die Zahlen?* [*op. cit.*]. He starts out from the existence of an *infinite* set and obtains by a coordination the set of natural numbers. But the existence of an infinite set is not presupposed by Zermelo.

Elsewhere too the concept of coordination is indispensable, for one can base upon it a general definition of the concept of a cardinal number. One namely ascertains that two finite sets are equinumerous[†] by numbering them both. But it is obviously a round-about path, that we should coordinate the set to be numbered first of all with the set of numbers. We shall rather try to see whether those sets can be coordinated with each other immediately, and if this is the case call them – following G. Cantor – equivalent. However, the importance of this direct method of likening comes to light when one considers that one cannot liken infinite sets in any other way at all, as was made clear by Cantor's work.

The concept of equivalence now also leads to a definition of finite cardinal number. Since the proposition holds that two sets equivalent to a third are also equivalent to each other, one will say of all sets equivalent to one another that they have the same "power" or – if they are finite – the same cardinal number. In this way G. Frege gives the definition (*Grundlagen der Arithmetik* ['The foundations of arithmetic'], Breslau, 1884, p. 79; see also *Grundgesetze der Arithmetik* ['The basic laws of arithmetic'], Jena, 1893, pp. 57–8, §§40–42): the cardinal number belonging to the concept F is the extension of the concept "equinumerous with the concept F". This definition is essentially taken over by B. Russell (*Principles of Mathematics*, Cambridge, 1903, pp. 111–16; A. N. Whitehead and B. Russell, *Principia Mathematica*, Cambridge, 1912, vol. II, p. 4): the cardinal number of a set is

---

[†] ['gleichzahlig' , i.e. numerically alike.]

the set of all sets equivalent to it. Thus e.g. the number of fingers on one hand is the set of all the sets which can be one-one coordinated with the fingers on one hand[†].

It is to be noted, however, that in this definition Russell relies upon a preceding more careful definition of the concept of a set. Namely in order to avoid the paradoxes of set theory, he has substituted for Frege's logical theory of arithmetic another one (the theory of logical types), in which the concept of a set automatically gets a narrower significance[*].

[10] See also E. Schröder, *Lehrbuch der Arithmetik und Algebra* [*op. cit.*], p. 15.

[11] This law is called the distributive law. The associative and commutative laws of addition and multiplication, and besides them the distributive law, are also satisfied by the ordinary system of complex numbers, moreover all of them except the commutative law of multiplication are satisfied by the system of quaternions (note 28).

Lastly, quite similar laws hold in the calculus of logic. If by the sum of the extensions of two concepts a and b one understands the extension of whatever is a or b ($a=$man, $b=$woman, $a+b=$human being)[††], and by ab the extension of whatever falls under both a and b ($a=$red, $b=$rose, $ab=$red rose), there then hold:

$$(a+b)+c=a+(b+c)$$
$$a+b=b+a$$
$$(ab)\,c=a(bc)$$
$$ab=ba$$
$$(a+b)\,c=ac+bc$$
$$ab+c=(a+c)\,(b+c)$$

There are thus *two* distributive laws.

[12] Of course, a concept too can in the last analysis be made understandable to a listener only by display, but in a different manner from e.g. a length.

One might think, in principle, that there would exist transitions between all the types of things in nature. But it is not so. One can find no continuous transitions from certain types to others. It is therefore possible to renew acquaintance with such a type in memory, after becoming acquainted with it upon a single occasion, without having the original itself at one's disposal, without measuring. Thus we can renew acquaintance e.g. with water and the states of freezing and boiling, and also arrive thereby at a conceptual definition of degrees of temperature. At present, the units of length, mass and time cannot be defined in such a manner.

[†] [Currently popular versions of set theory prefer not to use such a 'set of all sets' as the appropriate cardinal number, but instead an abstractly defined representative member of that set. (One can then avoid the complications of Russell's type theory, mentioned below.) Then e.g. the number five becomes an abstractly defined perfect example of five objects. Such a theory of number is not inaptly termed 'Platonism' .]

[*] e.g. one is not allowed to consider sets whose elements are partly simple things and partly sets. According to Russell, moreover, assertions about sets are nothing but assertions about things having certain properties.

[††] [This example is misleading, inasmuch as man and woman are exclusive concepts; the laws also hold e.g. for $a=$man and $b=$doctor, when $(a+b)$ covers whatever is a man or a doctor *or* both.]

We shall see, however, that this view must perhaps be modified (note 30).

[13] One calls a relation aRb between a and b symmetric, if from aRb there also follows bRa; one calls it transitive if from aRb and bRc there also follows aRc. Alikeness is thus a symmetric-transitive relation. Conversely, in all cases where a symmetric-transitive relation exists one can express oneself, regarding the things having the relation, as if these possessed a common property (definition by abstraction: G. Peano, *Notations de logique mathématique* ['Notations of mathematical logic'], Turin, 1894, p. 45).

For example, there is the symmetric-transitive relation between point pairs which consists in the fact that the ends of a rod which coincide with the one pair, can also coincide with the other pair (Helmholtz, this edition, p. 42). One calls such pairs of extents alike in magnitude. Also, the symmetric-transitive relation between bodies which can come into contact without an exchange of heat taking place. We call such bodies alike in hotness, and say of them that they have the same temperature. In a corresponding manner, G. Frege and B. Russell have defined the concept of a cardinal number (note 9).

In all such cases the following surmises suggest themselves:

(a) the objects concerned actually have a property in common, from which that relation issues;

(b) they are *quantitatively* alike in something.

Thus e.g. the property of two bodies such that they give no occasion for heat exchange when in contact, is explained by their having a like amount of energy per degree of freedom.

Regarding surmises of the first kind, one may adopt this viewpoint: the possession of a common property indeed means nothing but the existence of a physical relationship between those bodies, involving possibly other bodies (e.g. alikeness of extents – a behaviour of bodies towards measuring rods).

In this sense, one can ascribe to the proof for Russell's principle of abstraction (*Principles*, note to p. 162; *Principia*, vol. I, p. 474, no. 72.66) a significance which is not merely formal: all things between which the symmetric-transitive relations exist and which together form the set M, have the property in common of being an element of the set M (somewhat differently expressed in Russell).

[14] Compare Leibniz: 'Eadem sunt, quorum unum potest substitui alteri salva veritate' ["Those are the same of which one can be substituted for the other with truth preserved"], in *Non inelegans specimen demonstrandi in abstractis* ['A not inelegant example of demonstration in abstract things'], from *Opera*, ed. Erdmann, part I, p. 94.

[15] The proportionality between inertial mass and gravitational mass was demonstrated by Newton and Bessel using pendulum experiments, and by Eötvos, Pekar and Fekete (see e.g. *Naturwissenschaften* **7** (1919), 326) using experiments with the torsion balance. Since the observed gravitational force is composed out of the attractive force of the earth and the centrifugal force, it would have to take different directions – if proportionality did not hold – according to the choice of body, and a horizontal torsion balance with different materials at its ends would have to turn. But the investigators mentioned found no such effect.

These results became of heightened interest through Einstein's discovery that according to the theory of relativity, each energy form must possess inertia. It was now to be anticipated that it would also possess gravitational mass – and for this too one could in turn find a theoretical basis.

The general theory of relativity enables us to discern clearly the ground of the connexion between inertial and gravitational mass. Indeed: in a gravitational field a heavy body will produce a reading on a spring balance; but one can deny the presence of a gravitational field, and assume instead of it an acceleration of the balance. One will then see in the reading of the spring balance a consequence of the inertia of the body. Inertial and gravitational mass must therefore be alike with each other.

This gravitational mass is also to be distinguished conceptually from the active gravitational mass – likewise proportional to it – which measures the magnitude of the gravitational effects emanating from it.

[16] See note 13.

[17] The surfaces are not themselves brought up to each other, but instead the rays emanating from them are directed in the photometer in such a way that they meet the eye as neighbouring ones (Helmholtz, *Physiologische Optik* [*op. cit.*], pp. 419–422).

According to E. Schrödinger (*Annalen der Physik* **63** (1920), 397, 427, 481), very slightly different lights of different colours count as alike in brightness if an alteration in intensity of one or the other increases the dissimilarity. Only under certain conditons, which Schrödinger has likewise laid down, can one speak of like brightness of finitely different lights of different colours.

[18] The differential galvanometer contains two coils wound in opposite directions, such that the currents flowing through them act upon the same magnetic needle. If the two currents have like intensity, they cancel out each others effects.

[19] It is impossible to denote real objects, conceptually or by display, in such a way that there would even be a sense to the question of whether their dimensions are in rational or irrational proportion to one another. Can one then say that there exist irrational proportions in nature? In the case, on the other hand, of distances (or extents) which issue from one another by a mathematical construction, one can decide in principle regarding any of them whether it is rational or irrational.

[20] The theories of Weierstrass, Cantor and Dedekind (at least as they are mostly understood today: see M. Pasch, *Einleitung in die Differential- und Integralrechnung* ['Introduction to differential and integral calculus'], Leipzig, 1882, p. 3; B. Russell, *Principles of Mathematics* [*op. cit.*], p. 286; L. Couturat, *Les Principes des Mathématiques* ['The principles of mathematics'], Paris, 1905, p. 85; Deutsch, *Philosophische Prinzipien der Mathematik* ['Philosophical principles of mathematics'], Leipzig, 1908, p. 90) do not see in this narrowing down a *specification* of an irrational magnitude which exists in itself, but instead its *definition* too. According to this, statements about irrational numbers are ultimately statements about infinite sequences from the set of whole numbers[†].

[21] In the motion of a system of free points between which only central forces act, there are conserved:

      (1) the energy
      (2) the velocity components of the centre of gravity
      (3) the areal velocities.

[†] [This formulation of the relationship between the two kinds of number is over-simplified; one constructs the irrational numbers from infinite sets or sequences of rational numbers, but the rational numbers from sets of ordered pairs of whole numbers.]

The areal velocities are defined as follows: one projects the system onto a plane, and joins a fixed point in this plane to the projections of the material points by moving lines. The areas traced out per second by these lines are to be multiplied by the masses of the corresponding material points and added. One thus obtains the areal velocity in the plane concerned.

[22] One should firstly compare in this context note 30, where the nature of the concept of dimensions is explained. Since an n-fold force is obtained by putting together n like forces, force is an additive magnitude according to Helmholtz, a linear magnitude according to P. du Bois-Reymond, and its value a magnitude number[†] according to T. Ehrenfest-Afanasjewa (*Math.-Naturw. Blätter* **8** (1905); *Math. Ann.* **77** (1916), 259–276).

Between magnitudes of this kind there may exist further equations containing further numbers, but numbers which must be replaced with different ones when we alter the unit of measurement for the additive magnitudes. The magnitudes appearing in the equations therefore also have dimensions, as e.g. the gravitational constant in the equation:

$$F = g \cdot \frac{m_1 m_2}{r^2} \; .$$

Such magnitudes are called by Helmholtz constants, by Ehrenfest-Afanasjewa parameters.

There are here still two cases to be distinguished:

(1) Only one single equation is involved. The constant concerned is then a universal constant, as e.g. the gravitational constant determined by Newton's law, and the universal gas constant.

(2) The equation differs according to the kind of substance or thing between whose properties the equation asserts a relation. The constant concerned is then a constant for the substance or thing (specific conductivity, individual gas constant, conductivity).

[23] P. du Bois-Reymond (*op. cit.*, pp. 44–5) defines linear magnitudes essentially by the following requirement: two or more magnitudes of the same kind yield, when conjoined, once again a magnitude of the same kind, which is greater than its components.... if one magnitude is greater than another, then there is always a third of the same kind as those two, such that the second united with it yields the first. Forces are linear magnitudes (pp. 24, 28), as are the physical pitches of tones (p. 34) and sensations of intensity (p. 29). As examples of non-linear magnitudes one has complex numbers (p. 39), colour tones and perceived pitches of tones (p. 34).

It is evident that force will be called linear, since several forces can be compounded into a single one. Yet even from two colours one can compose a new colour (see e.g. note 29). But the components are in no way contained in the perceived colour tone. For this reason perceived colour tones are non-linear magnitudes. Yet why does du Bois-Reymond then also call sensations of intensity linear magnitudes? After all, the weaker one is not really contained in the stronger. The reason is that in our representation we coordinate, due to frequent experiences, two sensations with a third one in such a way that in fact the first of the corresponding stimuli can *really* be compounded from the second and third (p. 29).

One sees that it does not suffice for calling a kind of magnitudes linear, according to du Bois-Reymond, that they can be coordinated with numbers in a unique manner. So one will also understand Helmholtz' terming constants or parameters non-linear. Indeed,

---

[†] ['Grössenzahl']

two bodies of like volume, like mass and like density yield, when united, a body of twice the volume and twice the mass, but the *same* density. Something similar holds for specific conductivity, but in the case of connexion in parallel it fails to hold for conductivity, though this was listed as a constant. However, Helmholtz did admit the relativity of the classification.

Now one will also understand why the concept of a non-linear magnitude was linked with that of an intensive one. Helmholtz may there perhaps have been thinking of density. Calling this an intensive magnitude in fact seems quite suitable; after all, the pressure of a gas too depends upon its density. On the other hand, it looks odd to us at first to term a force an extensive magnitude. But we can represent to ourselves the forces occurring in nature as volume forces. Then a force density, which is to be regarded as an intensive magnitude, attaches to every point, while the total force upon a body is obtained by integration and may therefore count as an extensive magnitude.

[24] One calls oriented magnitudes vectors. Then e.g. the vector sum is found geometrically by the well-known parallelogram construction, while analytically by adding the components. It moreover holds: that if a vector is yielded as the sum of two others for *one* decomposition into components, then this must also be the case for any other such decomposition. Thus the sum does not depend upon the coordinate system. The aim of the vector calculus, however, is to avoid reference to a coordinate system from the start.

[25] One coordinates with an angular velocity a line whose orientation is the axis of rotation, whose length is equal to the angle gone through per second, and which is so oriented that when seen from the end point of the vector, the rotation takes place in the same sense as does a rotation from the X-axis to the Y-axis when seen from the Z-axis. Thus in a mirror reflection of the coordinate system the vector reverses its orientation. Such vectors, which one calls *axial* vectors and amongst which the magnetic field strength also belongs, are best portrayed each by a portion of surface for which a sense of rotation is prescribed.

[26] The square roots of negative numbers, which appear in the solution of algebraic equations, formerly counted as impossible or imaginary. Despite this, one calculated with them. Wallis, Bueé, Caspar Wessel, I. R. Argand and Gauss have given for the expression

$$a + \sqrt{-b} = a + b\sqrt{-1} = a + bi,$$

which one calls a complex number, a geometrical portrayal. As the representative of the number, one considers a point whose x-coordinate is equal to the "real" part a and whose y-coordinate is equal to the "imaginary" part b. However, the complex number can also be portrayed by the connecting line from the origin to this point, by a vector.

All the same, the geometrical viewpoint is not requisite for demonstrating that complex numbers are objects free from contradiction. They are namely best defined, according to W. R. Hamilton, as number pairs; one naturally must stipulate the rules of calculation with such number pairs by means of special definitions.

In the context of the present paper, the addition of complex numbers is above all of relevance. We put

$$(a_1 + ib_2) + (a_2 + ib_2) = (a_1 + a_2) + i(b_1 + b_2).$$

That means: in order to obtain the sum of two complex numbers, one forms the parallelogram of the representative vectors, or displaces the one vector along the other. In this manner one obtains the vector, or the point, which corresponds to the sum.

Complex numbers are multiplied as ordinary numbers, with replacement of i^2 by -1.

[27] H. Grassmann's theory of extension appeared in 1844, and remained unnoticed for 23 years until in the year 1867 Hermann Hankel drew the attention of mathematicians to it. It contains the principles of the vector calculus, which has today become indispensable to the physicist.

[28] W. R. Hamilton, who also first gave an arithmetical interpretation for the complex numbers, sought for number systems put together from even more than two units. Quaternions was what he called numbers of the form

$$a_0 + a_1 i + a_2 j + a_3 k$$

which are added in the usual manner, while there hold for the units i, j, k the multiplicative laws:

$$i^2 = j^2 = k^2 = -1$$
$$ij = -ji = k$$
$$jk = -kj = i$$
$$ki = -ik = j.$$

The system of quaternions satisfies the associative and commutative laws of addition, the associative law of multiplication, and the distributive – but not the commutative – law of multiplication. Apart from the systems of the real and the complex numbers, it is the only one which satisfies the mentioned axioms and is so constituted that a product only vanishes when one of the factors vanishes (Frobenius' theorem).

Quaternions lacking the first term can be interpreted as vectors, and behave like these under addition.

[29] Newton, *Opticks*, book I, part II, proposition VI, problem II. Compare also Helmholtz, *Physiologische Optik* [*op. cit.*], pp. 326 ff.

The basis of the Newton-Helmholtz portrayal is the empirical law that mixtures which look alike yield like colours when mixed (H. Grassmann), just as we can replace a system of forces not merely as such with its resultant, but also for combinations.

We choose three basic colours, and coordinate with them three base points. If a mixture contains the portions m_1, m_2, m_3 of the basic colours, its representative point is the centre of gravity of the three masses m_1, m_2, m_3 placed each at a base point[†], and at that centre itself we put a mass $m_1 + m_2 + m_3$ (thus the mixture has the intensity $m_1 + m_2 + m_3$). From the fact that in determining centres of gravity any arbitrary point masses can be conjoined at their centre of gravity, and from the mentioned law of mixtures for colours, it now follows that the representative points of any mixtures whatsoever can be found in general as centres of gravity of the representative points of the components.

A colour is therewith characterized by a point P and a metrical number m. However, one can also choose a point O lying outside the triangle specified by the base points, and let the vector $m \cdot OP$ be representative of the colour. Thus if $x_1, ..., z_3$ are the coordinates

---

[†] [i.e. at the corresponding base point.]

of the base points when referred to a coordinate system with O as origin, and if the portions m_1, m_2, m_3 are mixed, then the representative centre of gravity gets the coordinates

$$\frac{m_1 x_1 + m_2 x_2 + m_3 x_3}{m_1 + m_2 + m_3} \qquad \frac{m_1 y_1 + m_2 y_2 + m_3 y_3}{m_1 + m_2 + m_3}$$

$$\frac{m_1 z_1 + m_2 z_2 + m_3 z_3}{m_1 + m_2 + m_3}.$$

Thus the end point of the vector just constructed gets the coordinates $m_1 x_1 + m_2 x_2 + m_3 x_3$, $m_1 y_1 + m_2 y_2 + m_3 y_3$, $m_1 z_1 + m_2 z_2 + m_3 z_3$; or the coordinates m_1, m_2, m_3 if referred to an oblique-angled coordinate system whose axes go through the base points and whose origin is at O.

Thus in the new portrayal, the vector corresponding to the mixture m_1, m_2, m_3 arises by multiplying the base vectors by m_1, m_2, m_3 and adding them (see note II.8).

[30] In this inventory, one terms the expression on the right hand side the dimensions. We shall elucidate this concept with the example of force. The concept of an n-fold force is initially defined not e.g. by the well-known equation.
defined not e.g. by the well-known equation

$$F = m \cdot \frac{d^2 l}{dt^2}$$

(F = force, m = mass of a material point, l = path covered, t = time), but thus: like forces are such as can replace or neutralize each other; the n-fold force corresponding to a given one is such as can replace or neutralize n of the given like forces.

It is then *found* that force is proportional to a mass (namely that of the material point), to a length and to the inverse square of a time. One therefore uses the above equation to define the unit of force (not the concept 'n-fold force'), which is not contradicted by the fact that F may also be proportional to other powers of other masses, lengths and times. Indeed, according to the law of gravitation force is proportional to an expression $m_1 \cdot m_2 / r^2$, where m_1 and m_2 are masses and r is their separation (see Ehrenfest-Afanasjewa, *op. cit.*).

If one defines the unit of force such that we have the equation

$$F = m \cdot \frac{d^2 l}{dt^2}$$

then it holds that: if the units of mass, length and time are altered such that the corresponding numbers measuring them are multiplied by μ, λ and τ, then the number measuring force is multiplied by $\mu \lambda / \tau^2$. The expression $m \cdot l / t^2$ is in this system therefore called the dimensions of force. But consideration of the example given above is enough to refute the view that physical equations, by their very nature, can only contain magnitudes such that both sides are dimensionally alike. Admittedly, this condition can be enforced by attributing dimensions to the numbers, parameters or coefficients (see Helmholtz, this edition pp. 99–100), which occur in these equations (Ehrenfest, *op. cit.*).

The importance of the concept of dimensions rests upon the circumstance that the units of mass, length and time are in fact something wholly arbitrary, and that we therefore have reason to ask about the consequences of altering them. It is in this respect characteristic that we customarily do not attribute dimensions to temperature and the magnitudes of angles. Nevertheless, magnitudes involving them (the gas constant, angular velocity) will be measured by different numbers when one chooses different units for them (Reaumur instead of Celsius, ordinary degrees instead of radians). What decides is the following: one can introduce a unit of measure for these once and for all by definition ($100° =$ temperature difference between ice and steam), or by construction (for the angle $1°$). Such a procedure seemed to be lacking for the construction of a unit length, for which reason one had to take refuge in the standard meter and standard kilogram.

It was later recognized, however, that one possesses unaltering lengths in the wavelengths of light. In this way A. A. Michelson (*Traveaux et mémoires du bureau des poids et mesures* **11** (1894)) evaluated the meter precisely in wavelengths of cadmium. Purely theoretically, one can, say, choose the radius of the electron. Thus there are lengths which can be defined in a conceptual manner (compare note 12).

Now Euclidean space is distinguished from spherical or elliptical space by the fact that in it all lengths have like status, that therefore lengths and angles play different roles even in the theorems of elementary geometry, and similar figures are possible. It is otherwise with elliptical geometry (note II.30, under 1): in each of its theorems straight lines and angles can be exchanged; there is a greatest length, there are no similar figures, and certain lengths can be reobtained by prescription (note I.17).

Might then a hint of the non-Euclidean character of space be contained in the fact that there are classes of bodies whose dimensions are invariably alike (electrons, hydrogen nuclei)? In this context we may also recall that the world is represented by Einstein as a finite space of elliptical metric. Should this view be confirmed, lengths would not need to be treated in fundamentally different ways from angles, and there would be no more necessity for giving dimensions to any magnitudes. At the same time, we would also understand why the various types of things are discontinuously distinct from one another.

Even if we disregard such speculations, however, many alterations could still be made in our metrical system. From the law of gravitation, mass could be expressed in terms of length and time. A theoretically possible definition for the units would also be: the radius of the electron, the mass of the electron, and the time needed by light to cross the radius of the electron. Admittedly, this radius is not known exactly. But with spectroscopic measurements one can ascertain fairly exactly (A. Sommerfeld, *Atombau und Spektrallinien* ["Atomic structure and spectral lines"], Brunswick, 1921, p. 367): the electronic charge, the electronic mass, and a constant h introduced by Planck into radiation and quantum theory. If one sets these three constants $= 1$, one thereby obtains indirectly the units of mass, length and time.

[31] See notes 28 and 32.

[32] The vector product of two vectors having the components a_x, a_y, a_z and b_x, b_y, b_z is an axial vector (notes 24, 25), which possesses the components

$$a_y b_z - a_z b_y$$
$$a_z b_x - a_x b_z$$
$$a_x b_y - a_y b_x.$$

The theory of quaternions too leads to these formulae, since the quaternions $a_x i + a_y j + a_z k$ and $b_x i + b_y j + b_z k$, when multiplied according to the law of multiplication (note 28), yield

$$(a_y b_z - a_z b_y) i + (a_z b_x - a_x b_z) j + (a_x b_y - a_y b_x) k .$$

CHAPTER IV

THE FACTS IN PERCEPTION[1]

Address given during the anniversary celebrations of the
Friedrich Wilhelm University in Berlin, in 1878; reprinted
in *Vorträge und Reden*, vol. II, pp. 215–247, 387–406.

My distinguished audience !

Today on the birthday of the founder of our university, the sorely-tried
King Friedrich Wilhelm III, we celebrate the anniversary of its foundation. The year of its foundation, 1810, fell in the period of the greatest
external stress upon our country. A considerable part of its territory had
been lost, the land was exhausted from the preceding war and the enemy
occupation. The martial pride which had remained with it from the
times of the great elector and the great king had been deeply humiliated.
And yet this same period now seems to us, when we glance backwards, to
have been so rich in possessions of a spiritual kind, in inspiration, energy,
ideal hopes and creative thoughts, that we might, despite the relatively
brilliant external situation of our country and nation today, look back
upon it almost with envy. If in that distressing situation the king's first
thought was of founding the university before other material claims, if
he then staked throne and life so as to entrust himself to the resolute
inspiration of the nation in the struggle against the conqueror, this all
shows how deeply within him too, the simple man disinclined to lively
expressions of feeling, acted a trust in the spiritual powers of his people.

At that time Germany could point to a magnificent series of praise-
worthy names in both art and science, names whose bearers are in part
to be counted amongst the greatest of all times and peoples in the history
of human culture.

Goethe was alive and so was Beethoven; Schiller, Kant, Herder and
Haydn had survived the first years of the century. Wilhelm von Humboldt
was outlining the new science of comparative linguistics; Niebuhr, Fr.
Aug. Wolf and Savigny were teaching how to permeate ancient history,
poetry and law with living understanding; Schleiermacher was seeking
a profound understanding of the spiritual content of religion. Joh. Gott-
lieb Fichte, the second rector of our university, the powerful and fearless
public speaker, was carrying his audience away with the stream of his
moral inspiration and the bold intellectual flight of his idealism.

Even the aberrations of this mentality, which express themselves in the easily recognizable weaknesses of romanticism, have something attractive compared with dry, calculating egoism. One marvelled at oneself in the fine feelings in which one knew how to revel, one sought to develop the art of having such feelings. One thought oneself allowed to admire fantasy all the more as a creative force, the more it had freed itself from the rules of the understanding. Much vanity lay hidden in this, but all the same a vanity of enthusiasm for high ideals.

The older ones amongst us still knew the men of that period, who had once entered the army as the first volunteers, always ready to immerse themselves in the discussion of metaphysical problems, well-read in the works of Germany's great poets, men who still glowed with rage when talking of the first Napoleon, but with rapture and pride when of deeds in the war of liberation.

How things have changed! We may well exclaim thus with amazement in a period when a cynical contempt for every ideal possession of humankind is propagated, on the streets and in the press, and has reached its peak in two revolting crimes[†], which were obviously only aimed at the head of our emperor because in him was united everything that humanity, up to now, has regarded as worthy of veneration and gratitude.

We must almost make an effort to recall that only eight years have passed since the great hour, when at the call of the same monarch every rank of our people, without hesitation and filled with self-sacrificing and inspired patriotism, went into a dangerous war against an opponent whose might and valour were not unknown to us. We must almost make an effort to take note of the wide extent to which the endeavours, political and humane, to give the poorer ranks too of our people an existence less troubled and more worthy of human beings, have captured the activity and thoughts of the educated classes. Or to think how much their lot in material and legal respects has actually been improved.

The nature of mankind seems simply to be such that next to much light one can always find much shadow. Political freedom initially gives the vulgar motives a greater licence to reveal themselves and to embolden each other, as long as they are not faced with a public opinion ready to offer energetic opposition. Even in the years before the war of liberation, when Fichte was preaching sermons calling upon his generation to re-

---

† [In 1878, two attempts were made to assassinate the emperor.]

pent, these elements were not lacking. He depicts conditions and senti-
ments as ruling which recall the worst of our times. "The present
age adopts in its basic principle a stance of haughtily looking down
upon those who, from a dream of virtue, let themselves be torn away
from pleasures, and rejoicing in the thought that one must get beyond
such things, and not at all be imposed upon in this manner."* The only
pleasure, going beyond the purely sensuous, which he concedes to be
known to the representatives of that age, is what he calls "delighting in
one's own artfulness". And yet, in this same period, there was being
prepared a mighty upswing which belongs to the most glorious events
in our history.

Although we therefore need not regard our period as beyond hope,
we should surely not soothe ourselves too easily with the consolation
that things were indeed not better in other times than now. It is nevertheless advisable, when such dubious processes are going on, that each
person should make a review – in the sphere given him to work in and
which he knows – of the situation of the work towards the eternal goals
of mankind: whether they are being kept in view, whether one has got
nearer to them. In the youthful days of our university science too was
youthfully bold and strong in hopes, its view was directed pre-eminently
towards the highest goals. Although these were not to be reached so
easily as that generation hoped, although it also emerged that long drawn
out particular labours had to prepare the path towards them, so that
initially the nature itself of the tasks demanded another kind of work
– less enthusiastic, less immediately directed towards the ideal goals –
it would still doubtless be pernicious should our generation have lost
sight of the eternal ideals of mankind, over and above subordinate and
practically useful tasks.

In that period, the fundamental problem placed at the beginning of all
science was the problem of epistemology: "What is true in our intuition
and thought?[2] In what sense do our representations correspond to actuality?"[3] Philosophy and natural science encounter this problem from
two opposite sides, it is a task common to both.

The former, which considers the mental side, seeks to separate out
from our knowledge and representation what originates from the influ-

* Fichte, *Werke*, vol. VII, p. 40.

ences of the corporeal world, in order to set forth unalloyed what belongs to the mind's own workings. Natural science, on the contrary, seeks to separate off that which is definition, symbolism, representational form or hypothesis, in order to have left over unalloyed what belongs to the world of actuality[†] whose laws it seeks. Both seek to execute the same separation, although each is interested in a different part of what is separated[4]. In the theory of sense perceptions, and in investigations into the fundamental principles of geometry, mechanics and physics, even the enquirer into nature cannot evade these questions. As my own studies have frequently entered both domains, I want to try to give you a survey of what has been done in this direction on the part of enquiry into nature.

Naturally, in the last analysis the laws of thought are no different in the man enquiring into nature from what they are in the man who philosophises. In all cases where the facts of daily experience – whose profusion is after all already very great – sufficed to give a percipient thinker, with an unconstrained feeling for truth, in some measure enough material for a correct judgement, the enquirer into nature must satisfy himself with acknowledging, that the methodically completed gathering of empirical facts simply confirms the result gained previously. But there also occur cases of the contrary kind. Such cases will justify the fact – if it needs to be justified at all – that in what follows the questions concerned are not everywhere given new answers, but to a great extent ones given long ago are repeated. Indeed, often enough even an old concept, measured against new facts, gets a more vivid illumination and a new look.

Shortly before the beginning of the present century, Kant had developed the doctrine of forms of intuiting and thinking given prior to all experience[5] – or (as he therefore termed them) "*transcendental*"[6] forms of intuiting and thinking – into which forms any content we may represent must necessarily be absorbed, if this content is to become a representation. Regarding the qualities of sensation, Locke had already established a claim for the share which our corporeal and mental makeup has in the manner in which things appear to us[7]. In this direction, investigations into the physiology of the senses, which were in particular completed and critically sifted by Johannes Müller and then summarized by him in the law of *specific energies of sensory nerves*, have now brought

[†] ['Wirklichkeit'; 'wirklich' has been translated consistently as 'actual' rather than 'real', because below Helmholtz expressly distinguishes it from 'reell'.]

the fullest confirmation, one can almost say to an unexpected degree. At the same time, they have thereby portrayed and made intuitive, in a very decisive and palpable manner, the essence and significance of such a subjective form, given in advance, of sensation. This theme has already often been discussed, for which reason my presentation of it today can be brief.

There occur two distinct degrees of difference between the various kinds of sensation. The more deeply incisive difference is that between sensations belonging to different senses, such as between blue, sweet, warm, highpitched: I have permitted myself to term this a difference in the *modality* of sensation. It is so incisive as to exclude any transition from the one to the other, any relationship of greater or lesser similarity. One cannot at all ask whether e.g. sweet is more similar to blue or to red. On the other hand, the second kind of difference – the less incisive – is that between different sensations of the same sense: I restrict the term a difference of *quality* to this difference alone. Fichte groups together these qualities of a single sense as a quality range[†], and terms a *difference of quality ranges* what I just called a difference of modality. Within each such range, transition and comparison are possible. We can make the transition from blue through violet and crimson into scarlet, and e.g. declare yellow to be more similar to orange than to blue.

What physiological investigations now show is that the deeply incisive difference does not depend, in any manner whatsoever, upon the kind of external impression whereby the sensation is excited, but is determined alone and exclusively by the sensory nerve upon which the impression impinges. Excitation of the optic nerve produces only light sensations, no matter whether objective light – i.e. aether vibrations – impinges upon it, or an electric current which we pass through the eye, or pressure on the eyeball, or straining of the nerve stem during rapid changes of the direction of vision. The sensation arising through the latter influences is so similar to that of objective light, that people for a long time believed in light actually developing in the eye. Johannes Müller showed that such a development does not on any account take place, that the sensation of light was indeed only there because the optic nerve was excited[8].

Just as on the one hand each sensory nerve, excited by however so

---

[†] ['Qualitätenkreis']

many influences, always gives only sensations from the quality range proper to itself, so on the other hand are produced by the same external influences – when they impinge upon different sensory nerves – the most varied kinds of sensation, these always being taken from the quality range of the nerve concerned. The same aether vibrations as are felt by the eye as light, are felt by the skin as heat. The same air vibrations as are felt by the skin as a quivering motion, are felt by the ear as a note. Here the difference in kind of the impression is moreover so great, that physicists felt at ease with the idea that agents as apparently different as light and radiant heat are alike in kind, and in part identical, only after the complete alikeness in kind of their physical behaviour had been established, by laborious experimental investigations in every direction.

But even within the quality range of each individual sense, where the kind of object exerting an influence at least codetermines the quality of the produced sensation, there still occur the most unexpected incongruities. In this respect, the comparison of eye and ear is instructive. For the objects of both – light and sound – are oscillatory motions[9], each of which excites different sensations according to the rapidity of vibration: in the eye different colours, in the ear different pitches.

If we allow ourselves, for the sake of greater perspicuity, to refer to the frequency relationships of light in terms of the musical intervals formed by corresponding tone frequencies, then the result is as follows: the ear is sensitive to some ten octaves of different tones, the eye only to a sixth[†], although the frequencies lying beyond these limits occur for both sound and light, and can be demonstrated physically. The eye has in its short scale only three mutually distinct basic sensations, out of which all of its qualities are composed by addition, namely red, green and bluish violet. These mix in sensation without interfering with one another[10]. The ear, on the other hand, distinguishes between an enormous number of tones of different pitches. No two chords composed out of different tones ring alike, while yet with the eye precisely the analogue of this is the case. For a white which looks alike can be produced with red and greenish blue from the spectrum, with yellow and ultramarine, with greenish yellow and violet, [with green, red and violet,][††] or with

---

[†] [i.e. the interval having this name, not a sixth of an octave.]
[††] [The concluding words of the sentence indicate that Hertz and Schlick omitted this phrase by mistake.]

any two or three – or with all – of these mixtures together. Were the situation alike with the ear, the consonance of C and F would sound like that of D and G, E and A, or C, D, E, F, G and A, etc. And – what is notable as regards the objective significance of colour – apart from the effect on the eye, one has not been able to detect a single physical connexion in which light which looked alike was regularly alike in value.

The whole foundation, finally, of the musical effect of consonance and dissonance depends upon the peculiar phenomenon of beats. The basis of these is a rapid alternation in intensity of tone, which arises from the fact that two tones almost alike in pitch alternately interact with their phases alike and opposed, and correspondingly excite now strong and now weak vibrations in a resonating body. The physical phenomenon might equally well occur through the interaction of two light-wave trains as through the interaction of two sound-wave trains. But the nerve must firstly be capable of being affected by both wave trains, and it must secondly be able to follow quickly enough the alternation of strong and weak intensity. The auditory nerve is markedly superior in the latter respect to the optic nerve. At the same time, each fibre of the auditory nerve is sensitive only to tones from a narrow interval of the scale, so that only tones situated quite near to each other in it can interact at all. Ones far from each other cannot interact, or not directly. When they do, this originates from accompanying overtones or combination tones. There therefore occurs with the ear this difference between resounding and non-resounding intervals, i.e. between consonance and dissonance. Each fibre of the optic nerve, on the other hand, is sensitive throughout the whole spectrum, although with different strength in different parts. Could the optic nerve at all follow in sensation the enormously rapid beats of light oscillations, then every mixed colour would act as a dissonance [11].

You can see how all these differences in the manner of action of light and sound are conditioned by the way in which the nervous apparatus reacts to them.

Our sensations are indeed effects produced in our organs by external causes [12]; and how such an effect expresses itself [13] naturally depends quite essentially upon the kind of apparatus upon which the effect is produced. Inasmuch as the quality of our sensation gives us a report of what is peculiar to the external influence by which it is excited, it may count as a symbol of it, but not as an *image*. For from an image one requires some kind of alikeness with the object of which it is an image – from a statue alikeness of form, from a drawing alikeness of perspective projection in the visual field, from a painting alikeness of colours as well. But a sign need not have any kind of similarity at all with what it is the sign of. The relation between the two of them is restricted to the fact that like objects exerting an influence under like circumstances evoke like signs, and that therefore unlike signs always correspond to unlike influences.

To popular opinion, which accepts in good faith that the images which our senses give us of things are wholly true[14], this residue of similarity acknowledged by us may seem very trivial. In fact it is not trivial. For with it one can still achieve something of the very greatest importance, namely forming an image of lawfulness in the processes of the actual world[15]. Every law of nature asserts that upon preconditions alike in a certain respect, there always follow consequences which are alike in a certain other respect. Since like things are indicated in our world of sensations by like signs, an equally regular sequence will also correspond in the domain of our sensations to the sequence of like effects by law of nature upon like causes.

If berries of a certain kind in ripening develop at the same time a red pigment and sugar, then a red colour and a sweet taste will always be found together in our sensation for berries of this type.

Thus although our sensations, as regards their quality, are only *signs* whose particular character depends wholly upon our own makeup, they are still not to be dismissed as a mere semblance, but they are precisely signs of *something*, be it something existing or happening, and – what is most important – they can form for us an image of the *law* of this thing which is happening.

So physiology too acknowledges the qualities of sensation to be a mere form of intuition[16]. But Kant went further. He spoke not only of the qualities of sensations as given by the peculiarities of our intuitive faculty, but also of space and time, since we cannot perceive anything in the external world without its happening at a specific time and being situated at a specific place. Specification in time is even an attribute of every internal perception as well. He therefore termed time the given and necessary *transcendental form of inner intuition*, and space the corre-

sponding form of *outer intuition*. Thus Kant considers spatial specifications too as belonging as little to the world of the actual – or to 'the
thing in itself' – as the colours which we see are attributes of bodies in
themselves, but [which]† are introduced by our eye into them.

Even here the approach of natural science can take the same path, up
to a certain limit. Suppose we namely ask whether there is a common
characteristic, perceivable in immediate sensation, whereby every perception relating to objects in space is characterized for us. Then we in
fact find such a characteristic in the circumstance that motion of our
body places us in different spatial relations to the perceived objects, and
thereby also alters the impression made by them upon us. But the impulse to motion, which we give through an innervation of our motor
nerves, is something immediately perceivable [17]. That we do something,
when we give such an impulse, is felt by us. What we do, we do not know
in an immediate manner. Only physiology teaches us that we put into
an excited state – or *innervate* – the motor nerves, that their stimulation
is passed on to the muscles, that these consequently contract and move
the limbs. Yet all the same we know, even without scientific study, which
perceivable effect follows each of the various innervations that we are
able to initiate.

That we learn it by frequently repeated attempts and observations,
may be demonstrated with assurance in a long series of cases. We can
learn even as adults to find the innervations needed for pronouncing the
letters of a foreign language, or for a particular kind of voice production
in singing. We can learn innervations for moving our ears, for squinting
with our eyes inwards or outwards, or even upwards and downwards,
and so on. The difficulty in performing such things consists only in our
having to seek, by making attempts, to find the as yet unknown innervations needed for such previously unexecuted movements. We ourselves,
moreover, know of these impulses in no other form, and through no
other definable feature, than precisely the fact that they produce the
intended observable effect. Thus this effect also alone serves to distinguish the various impulses in our representation [18].

Now when we give impulses of this sort (turning our gaze, moving our
hands, going back and forth), we find that the sensations belonging to

---

† [Something is wrong with Helmholtz' syntax here, but the sense is obvious.]

certain quality ranges (namely those relating to spatial objects) can thereby be altered; other psychic states of which we are conscious – memories, intentions, wishes, moods – cannot be altered at all. A thoroughgoing difference between the former and the latter is thereby laid down in immediate perception.

Thus if we desire to call the relationship which we alter in an immediate manner by the impulses of our will – what kind of relationship this is might moreover be still quite unknown to us – a *spatial* one, then perceptions of *psychic* activities do not enter into such a relationship at all. But probably all sensations of the outer senses must proceed subject to some kind of innervation or another, i.e. have some spatial specification[†19]. In this case space will also appear to us – imbued with the qualities of our sensations of movement – in a sensory manner, as that through which we move, through which we can gaze forth. Spatial intuition would therefore be in this sense a subjective *form of intuition*, like the sensory qualities red, sweet and cold[20]. Naturally, the sense of this would just as little be mere semblance for the former as for the latter, the place specified for a specific individual object is no mere *semblance*[21].

From this point of view, however, space would appear as the *necessary* form of outer intuition, because precisely what we perceive as having some spatial specification comprises for us the external world. We comprehend as the world of inner intuition, as the world of self-consciousness, that in which no spatial relation is to be perceived[22].

And space would be a *given form* of intuition, possessed *prior to all experience*, to the extent that its perception were connected with the possibility of motor impulses of the will the mental and corporeal capacity for which had to be given to us, by our makeup[23], before we could have spatial intuition[24].

It will hardly be a matter for doubt, that the characteristic which we have discussed, of altering during movement, is an attribute of all perceptions relating to spatial objects*. The question will need to be answered, on the other hand, as to whether every specific peculiarity of our spatial intuition is now to be derived from this source. To this end we

---

† ['räumlich bestimmt sein' , i.e. any such sensation has a feature which can be altered by our moving.]
* On the localization of sensations of internal organs, see Appendix I to this paper.

must consider what can be attained with the aid of the features of perception which have so far been discussed.

Let us try to put ourselves back into the position of a man without any experience. We must assume, in order to begin without spatial intuition, that such a man knows even the effects of his innervations only to the extent that he has learnt how, by remission of a first innervation or by execution of a second counterimpulse, he can put himself back into the state from which he has removed himself by the first impulse. As this mutual self-cancellation of different innervations is wholly independent of what is thereby perceived, the observer can find out, without yet having previously gained any understanding of the external world, how he has to do this.

Let the situation of the observer initially be that he is faced with an environment of objects at rest. This will make itself known to him in the first place by the fact, that as long as he gives no motor impulse his sensations remain unaltered. If he gives such an impulse (e.g. if he moves his eyes or hands, or steps forward), the sensations alter; and if he then, by remission or the appropriate counterimpulse, returns to the earlier state, all his sensations will again be the earlier ones[25].

Let us call the whole group of sensation aggregates which can be brought about during the period of time under discussion, by a certain specific and limited group of impulses of the will, the *presentables* for that period; and call *present*, on the other hand, the sensation aggregate from this group which happens to be being perceived. Then our observer is tied at this time to a certain range of presentables, but any individual one of which he can make present at any moment he wishes by executing the relevant movement. Each individual presentable from this group thereby appears to him as *enduring at every moment* of this period of time. He has observed it at every individual moment that he wanted to. The assertion that he would have been able to observe it also at any other intervening moment that he might have wanted to, is to be regarded as an inductive inference, drawn from the case of every moment at which a successful attempt was made to that of every moment whatsoever in the relevant period of time. Therefore, the representation of an *enduring existence of different things at the same time one beside another* can in this manner be acquired.

'One beside another' is a spatial description. But it is justified, since we have defined as 'spatial' the relationship altered by impulses of the will. One does not yet need to think of substantial things as what are here supposed to exist one beside another. 'To the right it is bright, to the left it is dark, in front there lies resistance but not behind' could for example be said at this stage of knowledge, with right and left being only names for certain eye movements, in front and behind for certain movements of the hands.

Now at other times the range of presentables, for the same group of impulses of the will, is going to be a different one. This range, with the individual which it contains, will thereby confront us as something given, as an 'objectum' [†]. Those alterations which we can produce and revoke by conscious impulses of the will, are distinct from ones which are not consequences of such impulses and cannot be eliminated by them. The latter specification is negative. Fichte's appropriate expression for this is that the 'I' is faced with a 'not-I' which exacts recognition [26].

In asking about the empirical conditions under which spatial intuition develops, we must in these considerations take account chiefly of touch, since the blind can develop spatial intuition [27] completely without the help of sight. Although for them space will not turn out to be filled up with objects in such richness and detail as for sighted persons, it yet seems most highly improbable that the foundations of spatial intuition for the two classes of person should be wholly different. If we ourselves attempt to make observations by touching, in the dark or with our eyes closed, then we may very well touch with one finger – or even with a pencil held in the hand like the surgeon with his probe – and still ascertain, in detail and with assurance, the corporeal form of the object present.

When wanting to find our way in the dark, we usually feel over larger objects with five or ten finger-tips simultaneously. We then obtain five to ten times as many reports in the same time as with one finger, and also use the fingers, like the tips of an open pair of dividers, for measuring magnitudes in the objects. All the same, the circumstance that we have an extended sensitive skin surface, with many sensitive points, recedes wholly into the background when touching things. What we are capable of ascertaining from the skin feeling by gently applying our

---

[†] [The Latin word means variously 'cast in the way', 'opposed', 'offered'.]

hand, say upon the face of a medal, is extraordinarily rough and scanty in comparison with what we discover by a groping motion, even if only with the point of a pencil. With sight this process becomes much more complicated, because of the fact that besides the most refinedly sensitive spot on the retina – its central fovea – which is as it were led all round the retinal image when we look at something, there also cooperate at the same time a great host of other sensitive points, in a much more fertile manner than is the case with touch.

By moving the touching finger along the objects, one comes to know the sequence in which their impressions offer themselves. This sequence shows itself to be independent of whether one touches with one finger or another. It is moreover not a uniquely determined sequence, whose elements one must always go through, forwards or backwards, in the same order in order to get from one to another; thus it is not a linear sequence, but a surfacelike 'one beside another', or in Riemann's terminology a second-order manifold. That all this is so is easily seen.

Of course, the touching finger can get from one point to another, in the touchable surface, also by other motor impulses than those which push it along the surface; and different touchable surfaces require different movements for sliding upon them. A higher manifold is thereby required for the space in which what touches moves, than for the touchable surface: the third dimension must be added. But this suffices for all available experiences. For a closed surface divides completely the space with which we are acquainted[28]. Even gases and liquids, which after all are not tied to the form of the human faculty of representation[29], cannot escape through a surface closed all round. And just as only a surface, not a space – thus a spatial structure of two and not three dimensions – can be bounded by a closed line, so also can a surface close off precisely only a space of three dimensions, and not one with four.

Thus might one get to know the spatial order of what exist 'one beside another'. As a further step, magnitudes would be likened with one another, by observing congruence of the touching hand with parts or points of the surfaces of bodies, or congruence of the retina with parts and points of the retinal image.

Because this intuited spatial order of things stems originally from the sequence in which the qualities of sensation offered themselves to the moving sense organ, there finally persists a curious consequence even in the completed representation of an experienced observer. The objects extant in space namely appear to us clothed in the qualities of our sensations. To us they appear red or green, cold or warm, to have a smell or taste, etc., whereas after all these qualities of sensation belong only to our nervous system and do not reach out at all into external space[30]. The semblance does not cease even when we know this, because in fact this semblance is the original truth: it is indeed sensations which first offer themselves to us in a spatial order[31].

You can see that the most essential features[32] of spatial intuition can in this way be derived. However, to the consciousness of the general public an intuition appears as something simply given, which comes about without reflection and search, and which is by no means to be resolved further into other psychic processes. This popular belief has been adopted by some workers in physiological optics, and also by the strictly observant Kantians, at least as regards spatial intuition. As is well known, already Kant assumed not only that the general form of spatial intuition is transcendentally given, but that it also contains in advance, and prior to any possible experience, certain narrower specifications as expressed in the axioms of geometry[33]. These can be reduced to the following propositions[34]:

(1) Between two points only *one* shortest line is possible. We call such a line "*straight*".

(2) Through any three points a *plane* can be placed. A plane is a surface which wholly includes any straight line if it coincides with two of its points.

(3) Through any point only one line parallel to a given straight line is possible. Two lines are *parallel* if they are straight lines lying in the same plane which do not intersect within any finite distance.

Indeed, Kant used the alleged fact that these geometrical propositions appeared to us as *necessarily* correct, and that we could never at all even represent to ourselves a deviating behaviour of space, directly as a proof that they had to be given prior to all experience, and that for this reason the spatial intuition contained in them was itself a transcendental[35] form of intuition, independent of experience.

In view of the controversies which have been conducted, in recent years, about the question of whether the axioms of geometry are transcendental or empirical propositions, I should like here to emphasize firstly that this question is wholly to be separated from the one first discussed, of whether space is in general a transcendental form of intuition or not*.

Everything our eye sees, it sees as an aggregate of coloured surfaces in the visual field – that is its form of intuition [36]. The particular colours which appear on this or that occasion, their arrangement and sequence – this is the result of external influences and is not determined by any law of our makeup. Similarly from the fact that space is a form of intuiting, nothing whatever follows about the facts expressed by the axioms. If such propositions are taken to be not empirical ones, but to belong instead to the necessary form of intuition, then this is a further particular specification of the general form of space; and those grounds which allowed the conclusion that the form of intuition of space is transcendental, do not necessarily for that reason already suffice to prove, at the same time, that the axioms too are of transcendental origin [37].

When Kant asserted that spatial relationships contradicting the axioms of Euclid could never in any way be represented, he was influenced by the contemporary states of development of mathematics and the physiology of the senses, just as he was thus influenced in his whole conception of intuition in general as a simple psychic process, incapable of further resolution [38].

If one wishes to try to represent to oneself something which has never before been seen, one must know how to depict to oneself the series of sense impressions which, according to the known laws of the latter, would have to come about if one observed that object and its gradual alterations successively from every possible viewpoint and with all of one's senses [39]. And at the same time, these impressions must be such that every other interpretation is thereby excluded [40]. If this series of sense impressions can be formulated completely and unambiguously, then one must in my judgement declare that thing to be intuitably representable. Since by presupposition it is a thing which is considered never yet to have been observed, no previous experience can come to our help and guide our fantasy in seeking out the requisite series of impressions; instead, this can only happen by way of the *concept* of the object or relationship to be represented. Such a concept is thus first of all to be elaborated and to be made as specialised as the given purpose requires.

---

* See Appendix II below.

The concepts of spatial structures which are taken not to correspond to customary intuition can be reliably developed only by means of calculative analytic geometry. The analytic resources for our present problem were first given by Gauss in 1828 with his essay on the curvature of surfaces, and applied by Riemann in seeking out the logically possible self-consistent systems of geometry. These investigations have not unsuitably been termed *metamathematical*[41]. One should also note that already in 1829 and 1840 Lobatschewsky worked out, in the customary synthetically intuitive manner, a geometry without the axiom of parallels, and one which concurs completely with the corresponding parts of the more recent analytic investigations. Finally, Beltrami has formulated a method for forming images of metamathematical spaces in parts of Euclidean space, by means of which the specification of their manner of appearance in perspective vision is made fairly easy. Lipschitz has demonstrated that the general principles of mechanics can be carried over to such spaces, so that the series of sense impressions which would come about in them can be completely formulated. With this the intuitability of such spaces, in the sense of the definition of this concept given above, has been shown*.

But here is where disagreement occurs. I demand for the proof of intuitability only that one should be able to formulate, for every manner of observation, specifically and unambiguously the arising sense impressions, by using if necessary a scientific acquaintance with their laws, from which[†] it would ensue, at least for someone acquainted with these laws, that the thing concerned or relationship to be intuited was in fact present[42]. The task of representing to oneself the spatial relationships in metamathematical spaces indeed demands some practice in understanding analytic methods, perspective constructions and optical phenomena.

This is however in disagreement with the older concept of intuition, which only acknowledges something to be given through intuition if its representation enters consciousness at once with the sense impression, and without deliberation and effort. Our attempts to represent mathematical[††] spaces indeed do not have the ease, rapidity and striking self-evidence with which we for example perceive the form of a room which

---

* See my lecture on the axioms of geometry.
† [Apparently meaning: 'from which sense impressions' .]
†† [Helmholtz presumably means 'metamathematical' .]

we enter for the first time, together with the arrangement and forms of the objects contained in it, the materials of which these consist, and much else as well. Thus if this kind of self-evidence were an originally given and necessary peculiarity of all intuition, we could not up to now assert the intuitability of such spaces.

Yet we are now confronted with a host of cases, on further reflection, which show that assurance and rapidity for the occurrence of specific representations with specific impressions can also be acquired – even when no such connexion is given by nature. One of the most striking examples of this kind is our understanding of our mother tongue. Its words are arbitrarily or accidentally chosen signs – every different language has different ones. Understanding of it is not inherited, since for a German child who was brought up amongst Frenchmen and has never heard German spoken, German is a foreign language. The child becomes acquainted with the meaning of the words and sentences only through examples of their use. In this process one cannot even make understandable to the child – until it understands the language – that the sounds it hears are supposed to be signs having a sense. Lastly, on growing up it understands these words and sentences without deliberation and effort, without knowing when, where and through what examples it learnt them, and it grasps the finest variations of their sense – often ones where attempts at logical definition only limp clumsily behind.

It will not be necessary for me to multiply examples of such processes – they abound richly enough in daily life. This is precisely the basis of art, and most clearly that of poetry and the graphic arts. The highest manner of intuiting, as we find it in an artist's view, is this kind of apprehension of a new type of stationary or mobile appearance of man and nature. When the traces of like kind which are left behind in our memory by often repeated perceptions reinforce one another, it is precisely the law-like which repeats itself most regularly in like manner, while the incidental fluctuation is erased away. For the devoted and attentive observer, there grows up in this way an intuitive image of the typical behaviour of the objects which have interested him, and he knows as little afterwards how it arose as the child can give an account of the examples whereby it became acquainted with the meanings of words. That the artist has beheld something true emerges from the fact that it seizes us too with a conviction of its truth, when he presents it to us in an example purified from accidental perturbations. He is superior to us, however, in having known how to cull it from everything accidental, and from every confusion arising in the onward rush of the world.

Thus much just to recall how active this psychic process is in our mental life, from the latter's lowest to its highest stages of development. In previous studies I characterised as *unconscious inferences* the connexions between representations which thereby occur – unconscious, inasmuch as their major premiss is formed from a series of experiences, each of which has long disappeared from our memory and also did not necessarily enter our consciousness formulated in words as a sentence, but only in the form of an observation of the senses. The new sense impression entering in present perception forms the minor premiss, to which there is applied the rule imprinted by the earlier observations[43]. More recently I have avoided the name "unconscious inferences", in order to escape confusion with the – as it seems to me – wholly unclear and unjustified conception thus named by Schopenhauer and his followers. Yet evidently we are dealing here with an elementary process lying at the foundation of everything properly termed thought, even though it still lacks critical sifting and completion of the individual steps, such as occurs in the scientific formation of concepts and inferences[44].

Thus as concerns firstly the issue of the origin of the axioms of geometry: the fact that the representation of metamathematical spatial relationships is not easy when experience is lacking, cannot be claimed as a ground against their intuitability. Moreover, the latter is completely demonstrable. Kant's proof for the transcendental nature of the axioms of geometry is thus inadequate. On the other hand, investigation of the empirical facts shows that the axioms of geometry, taken in the only sense in which one is allowed to apply them to the actual world, can be empirically tested and demonstrated, or even – if the case should arise – refuted*.

The memory vestiges of previous experiences also play a further and highly influential role in the observation of our visual field. A no longer completely inexperienced observer receives even without movement of his eyes – whether by momentary illumination from an electric discharge

* See my *Wissenschaftliche Abhandlungen*, vol. II, p. 640; an excerpt is given as Appendix III below.

or by deliberate rigid staring – a relatively rich image of the objects in front of him. Yet even an adult will still easily convince himself that this image becomes much richer, and especially much more precise, when he moves his glance around in the visual field, and thus employs that kind of spatial observation which I described earlier as the fundamental one. We are indeed also so used to letting our glance wander upon the objects we observe, that it requires a fair amount of practice before we succeed, for the purposes of experiments in physiological optics, in holding it fixed upon one point for a longish time without wavering.

In my works on physiological optics*, I have sought to explain how our acquaintance† with the visual field can be acquired by observation of the images during the movements of our eyes, provided only that there exists, between otherwise qualitatively alike retinal sensations, some or other perceptible difference corresponding to the difference between distinct places on the retina. Such a difference should be called a *local sign*, according to Lotze's terminology[45]; except that the fact that this sign is a local sign – i.e. corresponds to a difference of place and to which such difference – need not be known in advance.

Recent observations** have also reconfirmed that persons who were blind from youth onwards, and later regained their sight through an operation, could not at first distinguish by eye even between such simple forms as a circle and a square, until they had touched them.

Apart from this, physiological investigation teaches us that we can liken by visual estimation, in a relatively precise and assured manner, only such lines and angles in the visual field as can be brought, by normal eye motions, to form images in rapid succession at the same places on the retina. Indeed, we estimate much more assuredly the true magnitudes and distances of spatial objects situated not too far off, than the perspective ones, alternating according to viewpoint, in the visual field of the observer – although the former task concerns the three dimensions of space and is much more involved than the latter, which concerns only a surfacelike

---

* *Handbuch der Physiologischen Optik* ['Handbook of physiological optics']; *Vorträge über das Sehen der Menschen* ['Lectures on human sight'], in *Vorträge und Reden*, vol. I, pp. 85 and 265.
† ['Kenntnis' : note the end of the first paper in this collection, where Helmholtz contrasts knowledge ('Erkenntnis') of a conceptual connexion and an intuitive acquaintance deriving from frequent observation.]
** Dufour (Lausanne) in the *Bulletin de la Société médicale de la Suisse Romande*, 1876.

image. One of the greatest difficulties in drawing, as is well known, is to free oneself from the influence involuntarily exerted by our representation of the true magnitudes of the objects seen. Now it is precisely the situation described that we must expect, if our understanding of local signs was first acquired through experience. We can assuredly become acquainted with the alternating sensory signs for what remains objectively constant, much more easily than with those for what alternates according to every single movement of our body, as the perspective images do.

There is, none the less, a large number of physiologists whose view we may term *nativist*, as opposed to the *empiricist* view which I myself have tried to defend, and for whom this conception of an acquired acquaintance with the visual field appears unacceptable. This is due to their not having got clear about what after all lies before us so plainly in the example of speech, namely how much can be achieved by accumulated memory impressions. For this reason, a host of various attempts have been made to reduce at least some part of visual perception to an innate mechanism, in the sense that certain impressions of sensation are supposed to release certain ready-made spatial representations.

I have demonstrated in detail* that all hypotheses of this kind proposed to date are inadequate, because in the end one can still always come up with cases where our visual perception is in more precise agreement with actuality than those assumptions would yield. One is then forced to add the further hypothesis that the experience gained during movements can in the end overcome the innate intuition, and thus achieve *in opposition* to the latter what it is supposed by the empiricist hypothesis to achieve *without* such an obstacle.

The nativist hypotheses about our acquaintance with the visual field thus *firstly* do not explain anything, but simply assume the existence of the fact to be explained while at the same time rejecting the possibility of reducing this fact to definitely ascertained psychic processes, although they themselves still have to appeal to the latter in other cases. *Secondly*, the assumption of every nativist theory – that ready-made representations of objects are elicited through our organic mechanism – appears much more audacious and doubtful than the assumption of the empiricist

---

* See my *Handbuch der Physiologischen Optik* [*op. cit.*], 3. Abteilung, Leipzig, 1867.

theory, which is that only the non-understood material of sensations originates from external influences, while all representations are formed from it in accordance with the laws of thought.

Thirdly, the nativist assumptions are unnecessary. The only objection that one has been able to bring against the empiricist explanation, is the assurance with which many animals move when newly born or just after crawling out of the egg[46]. The less mentally endowed they are, the quicker they learn what they at all can learn. The narrower the paths are along which their thoughts must go, the more easily they find them. A newly born human child is extremely inept at seeing: it needs several days before learning to judge, from its visual images, in what direction it must turn its head in order to reach its mother's breast. Young animals are certainly much more independent of individual experience. But we still know practically nothing specific about what this instinct which guides them is: whether direct inheritance is possible of ranges of representations from the parents, or whether it is only a matter of desire or aversion – or of a motor impulse – which attach themselves to certain sensation aggregates. Vestiges of the last mentioned phenomena still occur in a plainly recognisable manner with human beings. Properly and critically executed observations would be in the highest degree desirable in this domain.

Thus for the kind of set-up presupposed by the nativist hypothesis, one can at most claim a certain pedagogical merit which facilitates detection of the first lawlike relationships. The empiricist view too could be combined with presuppositions having this aim, for example that the local signs of neighbouring places on the retina are more similar to each other than are those of ones further apart, that those of corresponding places on the two retinas are more similar than those of disparate ones, and so on. For our present investigation it suffices to know that spatial intuition can come fully into being even with a blind person, and that with a sighted person – even should the nativist hypotheses prove partially correct – the final and most precise specification of spatial relationships is still conditioned by the observations made during movement[47].

I shall now return to our discussion of the initial, original facts of our perception. As has been seen, we do not merely have alternating sense impressions which come upon us without our doing anything about it. We rather observe during our own continuing activity, and thereby attain

an acquaintance with the *enduring existence* of a lawlike relationship
between our innervations and the becoming present of the various impressions from the current range of presentables. Each of our voluntary
movements, whereby we modify the manner of appearance of the objects, is to be regarded as an experiment through which we test whether
we have correctly apprehended the lawlike behaviour of the appearance
before us, i.e. correctly apprehended the latter's[†] presupposed enduring
existence in a specific spatial arrangement.

The chief reason, however, why the power of any experiment to convince is so much greater than that of observing a process going on without
our assistance, is that with the experiment the chain of causes runs
through our own self-consciousness. We are acquainted with one member
of [the chain of] these causes – the impulse of our will – from inner intuition, and know through what motives it came about[48]. From this, as
from an initial member known to us and at a point in time known to us,
there then begins to act that chain of physical causes which terminates in
the outcome of the experiment. Yet the conviction to be attained has an
essential presupposition, that the impulse of our will should neither
itself already have been influenced by physical causes which at the same
time also determined the physical process, nor should it for its own part
have influenced the subsequent perceptions psychically.

The latter doubt can in particular be of relevance to our present topic.
The impulse of the will for a specific movement is a psychic act, just as
is the thereupon perceived alteration in the sensation. Could not then
the first act bring about the second through purely psychic mediation?

It is not impossible. When we dream, something of the sort occurs. In
our dream we believe ourselves to execute a movement, and we then
dream further that there occurs what should be its natural consequence.
We dream of climbing into a boat, pushing it off from land, gliding out on
the water, seeing the displacement of the surrounding objects, and so
on. Here the dreamer's expectation that he will see the consequences of
his conduct occur appears to bring about the dreamed perception in a
purely psychic way. Who can say how long and finely spun out, how
logically complete such a dream might be? Should everything in it occur
in the most lawlike manner according to the order of nature, there would

---

† [or possibly 'their' , referring to 'the objects' rather than 'the appearance' .]

remain but one difference from the waking state – the possibility of being awakened, the rupture of this dreamed series of intuitions.

I do not see how we could refute a system of even the most extreme subjective idealism, if it regarded life as a dream[49]. We might declare it to be as improbable and unsatisfying as possible – in this respect I would assent to the sharpest expressions of repudiation. But it could be implemented consistently, and it seems to me very important to keep this in view. It is well known how ingeniously Calderón implemented this theme in his 'Life a Dream'.

Fichte too assumes that the 'I' posits the 'not-I' – i.e. the world as it appears – for itself, because it needs it for developing its own thought-activity. Yet his idealism does indeed distinguish itself from that just referred to, in that he conceives of other human individuals not as dream images, but – starting from the assertion of the moral law – as essences alike with one's own 'I'[50]. Since, however, the images whereby they each represent the 'non-I' must themselves all agree with one another, he conceives of all of the individual 'I's' as parts or emanations of the absolute 'I'. The world in which they found themselves was then that world of representations which the worldmind posited for itself, and could again receive the concept of reality, as happened with Hegel.

The *realist* hypothesis, on the other hand, trusts the testimony of ordinary self-observation, according to which the alterations of perception which follow some item of conduct have no psychic connexion at all with the preceding impulse of the will. It regards as enduring, independent of the way in which we form representations[†], that which seems to prove to be thus in everyday perception – the material world outside us.

The realist hypothesis is the simplest we can form, it has been tested and confirmed in extraordinarily wide ranges of application, it is sharply defined for every individual specification, and it is therefore extraordinarily serviceable and fruitful as a basis for conduct. All this is without doubt. Even in the idealist manner of conceiving things, we would hardly know how else to express the lawlike in our sensations than by saying: "Those acts of consciousness which occur with the character of perception proceed *as if* there actually existed the world of material things which is assumed by the realist hypothesis."[51] But we cannot get beyond this 'as if'. We cannot acknowledge the realist view to be more than an excellently serviceable and precise hypothesis. We are not allowed to

† ['von unserem Vorstellen'.]

ascribe to it necessary truth, since besides it idealist hypotheses not open to refutation are also possible.

It is good to keep this always before our eyes, so that we may not wish to infer more from the facts than there is to infer from them. The various gradations of the idealist and realist views are metaphysical hypotheses. As long as they are acknowledged to be such, they are ones which have their complete scientific justification, however harmful they may become when one wishes to present them as dogmas or alleged necessities of thought.

Science must discuss all admissible hypotheses, in order to retain a fully comprehensive view of the possible attempts at explanation. Hypotheses are even more necessary for conduct, because one cannot continually wait until an assured scientific decision has been reached, but must decide for oneself – whether according to probability or to aesthetic or moral feeling. In this sense one could have no objection even against metaphysical hypotheses. But it is unworthy of a thinker wishing to be scientific if he forgets the hypothetical origin of his propositions. When such concealed hypotheses are defended with pride and passionateness, the latter are the customary consequences of the unsatisfying feeling which their defender shelters, in the hidden depths of his conscience, about the justness of his cause.

Yet what we can find unambiguously, and as a fact without anything being insinuated hypothetically, is the lawlike in the phenomena. From the first step onwards, when we perceive before us the objects distributed in space, this perception is the acknowledgement of a lawlike connexion between our movements and the therewith occurring sensations. Thus even the first elementary representations contain intrinsically some thinking, and proceed according to the laws of thought. Everything in intuition which is an addition to the raw material of sensations can be resolved into thinking, if we take the concept of thinking as broadly as has been done above[52].

For if 'comprehending' means forming *concepts*[†], and in the concept of a class of objects we gather together and bind together whatever like characteristics they bear, it then results quite analogously, that the concept of a series of appearances alternating in time must seek to bind to-

---

[†] [Here there is a play on the words 'begreifen' and 'Begriff' which is untranslatable, as the former means more than merely 'conceiving' .]

gether that which remains alike in all of its stages [53]. The wise man, as Schiller puts it:

> Seeks the familiar law
>> in chance's frightful miracles,
> Seeks the stationary pole
>> in the fleeting appearances.[†]

That which remains alike, without dependence upon anything else, through every alternation of time, we call *substance*. The relationship which remains alike between altering magnitudes, we call the *law* connecting them. What we perceive directly is only this law [54]. The concept of substance can be gained only through exhaustive examination and always remains problematic, inasmuch as further examination is not ruled out. Formerly light and heat were counted as substances, until it later turned out that they are perishable forms of motion. And we must still always be prepared for new decompositions of the currently familiar chemical elements.

The first product of the thoughtful comprehension of the phenomena is lawlikeness. Should we have separated it out sufficiently purely, delimited its conditions with sufficient completeness and assurance and also formulated them with sufficient generality that the outcome is unambiguously specified for all possibly occurring cases, and that we at the same time gain the conviction that it has proved true and will prove true at all times and in every case: then we acknowledge it as an existence enduring independent of the way in which we form representations, and call it the *cause*, i.e. that which primarily remains and endures behind what alternates. In my opinion, only the application of the word in this sense is justified, although it is applied in common speech in a very wishy-washy manner for whatever at all is the antecedent or occasion of something[55].

Inasmuch as we then acknowledge the law to be something compelling our perception and the course of natural processes, to be a power equivalent to our will, we call it 'force'[56]. This concept of a power con-

---

[†] Sucht das vertraute Gesetz
    in des Zufalls grausenden Wundern,
  Suchet den ruhenden Pol
    in der Erscheinungen Flucht.
[The contexts show that Helmholtz intends this and subsequent snatches of poetry in their most literal sense. So the relevant literal sense is given in this translation.]

fronting us is conditioned directly by the way in which our simplest perceptions come about. From the beginning, those changes which we make ourselves by acts of our will are distinct from ones which cannot be made by our will, and are not to be set aside by our will. Pain especially gives us the most penetrating lesson about the power of reality. Emphasis thereby falls upon the observational fact that the perceived range of presentables is not posited by a *conscious* act of our representation or will. Fichte's 'not-I' is here the exactly fitting negative expression [57]. For the dreamer too, what he believes himself to see and feel appears not to be evoked by his will or by conscious concatenation of his representations, although the latter might in actuality often enough be unconsciously the case. For him too it is a 'not-I'. Likewise for the idealist, who regards it as the world-mind's world of representations.

In our language, we have a very fortunate way of characterising that which lies behind the change of appearances and acts upon us, namely as 'the actual'. Here only action[†] is predicated. Absent is that secondary reference to what endures as substance which is included in the concept of the real, i.e. of the thinglike. As regards the concept of the objective, on the other hand, the concept of a ready-made image of an object usually finds its way into it, and one which does not suit the most primary perceptions.

Even with the logical dreamer, we must presumably characterise as active and actual those psychic states or motives which foist upon him, at the given time, the sensations corresponding in a lawlike manner to the present situation in his dream world. It is clear, on the other hand, that a division between what is thought and what is actual does not become possible until we know how to make the division between what the 'I' can and cannot alter. This does not become possible, however, until we discern what lawlike consequences the impulses of our will have at the given time. The lawlike is therefore the essential presupposition for the character of the actual.

I need not explain to you that it is a *contradictio in adjecto* to want to represent the real, or Kant's 'thing in itself', in positive terms but

---

[†] [Here (and as far as possible elsewhere) the following equivalences are used: wirklich = actual, Wirken = action, (ein)wirken = to act, wirksam = active, reell = real, sachlich = thinglike, objektiv = objective, Gegenstand = object. Elsewhere 'Wirkung' is generally translated by 'effect' and 'Einwirkung' by 'influence'.]

without absorbing it into the form of our manner of representation. This is often discussed. What we can attain, however, is an acquaintance with the lawlike order in the realm of the actual, admittedly only as portrayed in the sign system of our sense impressions:

> Everything perishable
> Is only a likeness.[†]

I take it as a favourable sign that we find Goethe, here and further on, together with us on the same path. Where it is a matter of broad panoramas, we may well trust his clear and unconstrained eye for truth. He demanded from science that it should be only an artistic arrangement of the facts and form no abstract concepts going beyond this, which to him seemed to be empty names and only obscured the facts. In somewhat the same sense, Gustav Kirchhoff has recently characterised it as the task of the most abstract amongst the natural sciences, namely mechanics, to *describe completely and in the simplest manner* the motions occurring in nature[58].

As for 'obscuring', this indeed happens when we stay put in the realm of abstract concepts and do not explain to ourselves their factual sense, i.e. make clear to ourselves what observable new lawlike relationships between the appearances follow from them. Every correctly formed hypothesis sets forth, as regards its factual sense, a more general law of the appearances than we have until now directly observed – it is an attempt to ascend to something more and more generally and inclusively lawlike. Whatever factually new things it asserts must be tested and confirmed by observation and experiment. Hypotheses not having such a factual sense, or which in no way specify anything sure and unambiguous about the facts falling under them, are to be regarded only as worthless talk.

Every reduction of the appearances to the underlying substances and forces claims to have found something unchanging and final. An unconditional claim of this kind is something for which we never have a justification: this is allowed neither by the fact that our knowledge is full of gaps, nor by the nature of the inductive inferences upon which all of our perception of the actual, from the first step onwards, is based.

[†] Alles Vergängliche
Ist nur ein Gleichnis.

Every inductive inference is based on trusting that an item of lawlike behaviour, which has been observed up to now, will also prove true in all cases which have not yet come under observation. This is a trust in the lawlikeness of everything that happens. However, lawlikeness is the condition of comprehensibility. Trust in lawlikeness is thus at the same time trust in the comprehensibility of the appearances of nature. While: should we presuppose that this comprehension will come to completion, that we shall be able to set forth something ultimate and finally unalterable as *the cause* of the observed alterations, then we call the regulative principle of our thought which impels us to this the *law of causality*[†]. We can say that it expresses a trust in the *complete comprehensibility* of the world.

Comprehension, in the sense in which I have described it, is the method whereby our thought masters the world, orders the facts and determines the future in advance. It is its right and its duty to extend the application of this method to everything that occurs, and it has already actually harvested great yields on this path. However, we have no further guarantee for the applicability of the law of causality than this law's success. We could live in a world in which every atom was different from every other one, and where there was nothing at rest. Then there would be no regularity of any kind to be found, and our thought activity would have to be at a standstill[59].

The law of causality actually is an *a priori* given, a transcendental law[60]. A proof of it from experience is not possible, since the first steps of experience, as we have seen, are not possible without employing inductive inferences, i.e. without the law of causality. But even suppose that complete experience could tell us – though we are still far from being entitled to affirm this – that everything so far observed had occurred in a lawlike manner. It would still follow from such experience only by an inductive inference, i.e. by presupposing the law of causality, that the law of causality would then also hold in the future. Here the only valid advice is: have trust and act!

The inadequate
It then takes place.[††]

That should perhaps be the answer given by us to the question: what

[†] ['Kausalgesetz']
[††] Das unzulängliche
Dann wird's Ereignis.

is true in the way in which we form representations? Regarding what has always seemed to me to be the most essential advance in Kant's philosophy, we still stand on the ground of his system. In this regard I have also frequently emphasized in my previous studies the agreement[61] between recent physiology of the senses and Kant's doctrines, although this admittedly does not mean that I had to swear by the master's words in all subordinate matters too. I believe the resolution of the concept of intuition into the elementary processes of thought as the most essential advance in the recent period. This resolution is still absent in Kant, which is something that then also conditions his conception of the axioms of geometry as transcendental propositions. Here it was especially the physiological investigations on sense perceptions which led us to the ultimate elementary processes of cognition. These processes had to remain still unformulable in words, and unknown and inaccessible to philosophy, as long as the latter investigated only cognitions finding their expression in language.

Admittedly – for those philosophers who have retained the inclination for metaphysical speculations, the most essential thing in Kant's philosophy appears to be precisely what we have considered to be a defect hanging upon inadequate development in the specialised sciences of his time. Kant's proof, indeed, for the possibility of a metaphysics – and of course he himself did not know how to discover anything more about this alleged science – relies purely and simply on the belief that the axioms of geometry, and the related principles of mechanics, are transcendental propositions given *a priori*. Moreover his whole system properly speaking contradicts the existence of metaphysics, and the obscure points of his epistemology, about whose interpretation there has been so much controversy, derive from this root.

According to all this, science would seem to have its own secure territory standing firmly upon which it can seek for the laws of the actual – a marvellously rich and fruitful field of work. As long as it restricts itself to this activity, it will be unaffected by idealist doubts. Such work may appear modest in comparison with the soaring schemes of metaphysicians.

> Yet with gods
> Shall measure himself

No mortal.
If he raises himself up
And touches the stars
With the crown of his head,
Then nowhere cling
His uncertain soles,
And there play with him
Clouds and winds.

If he stands with firm
Pithy bones
On the well-founded
Lasting Earth:
In height he does not reach
Even with the oak
Or the vine
To liken himself.[†]

All the same, the example of the man who said this may teach us how a mortal, who had surely learnt how to stand, even when he touched the stars with the crown of his head, still retained a clear eye for truth and actuality. Something of the artist's vision, of the vision which led Goethe and also Leonardo da Vinci to great scientific thoughts, is what the genuine enquirer must always have. Both artist and enquirer strive, although with different approaches, towards the goal of discovering new lawlikeness. Only, one must not try to pass off idle enthusiasm and crazy fantasies as an artist's vision. The genuine artist and the genuine enquirer both know how to work genuinely, and to give their works a firm form and a convincing fidelity to truth.

[†] Doch mit Göttern
  Soll sich nicht messen
  Irgendein Mensch.
  Hebt er sich aufwärts
  Und berührt
  Mit dem Scheitel die Sterne,
  Nirgends haften dann
  Die unsicheren Sohlen,
  Und mit ihm spielen
  Wolken und Winde.

Steht er mit festen
Markigen Knochen
Auf der wohlgegründeten
Deuernden Erde:
Reicht er nicht auf,
Nur mit der Eiche
Oder der Rebe
Sich zu vergleichen.

Moreover, actuality has so far always revealed itself much more sub-
limely and richly to a science enquiring in a manner faithful to its laws,
than the utmost efforts of mythical fantasy and metaphysical speculation
had known how to depict it. What have all the monstrous offspring of
Indian reverie, the piling up of gigantic dimensions and numbers, to say
as against the actuality of the structure of the universe, as against the
intervals of time in which sun and earth were formed, in which life
evolved during geological history and adapted itself, in more and more
perfect forms, to the more stable physical situations on our planet?

What metaphysics has prepared, in advance, concepts of effects such
as magnets and moving electricity exert on each other? Physics at this
moment is still striving to reduce them to well-specified elementary
effects, without having reached a clear conclusion. But already light too
seems to be nothing other than one more kind of motion of these two
agencies, and the aether filling space is acquiring wholly new characteristic properties as a magnetisable and electrifiable medium.

And into what scheme of scholastic concepts shall we insert that supply
of effective energy whose constancy the law of the conservation of force
asserts? This supply, like a substance, cannot be destroyed or increased;
it is at work as a driving force in every motion of both lifeless and living
matter; it is a Proteus ever attiring itself in new forms, active throughout
infinite space and yet not divisible by space without remainder, the effective factor in every effect, the motive factor in every motion – and yet
not mind and not matter. Did the poet have a presentiment of it?

> In life's currents, in a storm of deeds,
> I float up and down,
> Weave here and there!
> Birth and grave,
> An eternal ocean,
> An alternating weaving,
> A glowing life,
> Thus I create on the humming loom of time
> And make the deity's living garment.[†]

[†] In Lebensfluten, in Tatensturm,   Ein ewiges Meer,
    Wall' ich auf und ab,   Ein wechselnd Weben,
    Webe hin und her!   Ein glühend Leben,
    Geburt und Grab,   So schaff' ich am sausenden Webstuhl der Zeit,
                                      Und wirke der Gottheit lebendiges Kleid.

We are particles of dust on the surface of our planet, which itself is barely to be called a grain of sand in the infinite space of the universe. We are the youngest generation of living things on earth, by the geological reckoning of time barely arisen from the cradle, still at the stage of learning, barely half-educated, and pronounced of age only out of mutual considerateness. Yet we have already – through the more powerful impulse of the law of causality – grown out above all of our fellow-creatures and are subduing them in the struggle for existence. We truly have sufficient ground to be proud that it has been given to us slowly to learn to understand, by faithful labour, "the incomprehensibly high works". And we need not feel in the least ashamed of not succeeding in this immediately at the first assault of a flight like that of Icarus.

APPENDICES

1. ON THE LOCALISATION OF THE SENSATIONS
OF INTERNAL ORGANS

The issue might arise here of whether the physiological and pathological sensations of internal organs of the body should not fall into the same category as psychic states, inasmuch as many of them are likewise not altered by movements, or at least not altered considerably[62].

Now there are indeed sensations of an ambiguous character, such as those of depression, melancholy and anxiety, which may just as well arise from bodily causes as from psychic ones, and for which there is also lacking any representation of a particular localisation[63]. At most, in the case of anxiety, the region of the heart vaguely asserts a claim to be the seat of the sensation, as in general the older view making the heart the seat of many psychic feelings was obviously derived from the fact that the movement of this organ is often altered by such feelings, a movement which one feels partly directly and partly indirectly through superimposing one's hand. So there thus arises a kind of false bodily localisation for what are actually psychic states. In states of illness this goes much further. I recall having seen, as a young doctor, a melancholic shoemaker who believed he could feel that his conscience had squeezed itself between his heart and his stomach.

On the other hand, there are of course a series of bodily sensations, such as hunger, thirst, satiety, neuralgic and inflammatory pains, which we localise as bodily ones – although uncertainly – and do not hold to be psychic, even though they are hardly altered by movements of the body. Admittedly, most inflammatory and rheumatic pains are considerably increased by pressure on the parts [of the body] in which they have their seat, or by movement of those parts. But even in the contrary case, as likewise neuralgic pains, they are probably to be regarded only as higher intensities of normally occurring feelings of pressure and strain on the parts concerned.

In this matter, the kind of localisation often gives an indication of what occasioned our learning something about the place of the sensation. Thus almost all sensations of the abdominal viscera are transferred to particular places on the anterior abdominal wall, even for organs such as the duodenum, pancreas, spleen and so on, which lie nearer to the posterior wall of the trunk. But pressure from outside can reach all of these organs almost only through the pliable anterior abdominal wall, not through the thick layers of muscle between the ribs, spine and hip-bone. It is further very noteworthy, that with toothache from an inflammation of the periosteum of a tooth the patients are usually uncertain at first which tooth is hurting – the upper or the lower – from a pair of teeth one above the other. One must first press forcefully upon the two teeth in order to find which is giving the pain. Would the origin of this not be that pressure on the periosteum of the root of the tooth in the normal state tends to occur only during chewing, and that both teeth of each pair then always suffer simultaneously a pressure of like strength?

The feeling of satiety is a sensation of fullness of the stomach, which is distinctly increased by pressure on the pit of the stomach, whereas the feeling of hunger is somewhat reduced by the same pressure. This may be what occasions their localisation in the pit of the stomach. Besides which, if we assume that like local signs accrue to nerves ending at the same places in the body, then the clear-cut localisation of one sensation of such an organ would suffice for its other sensations as well.

This probably holds for thirst too, to the extent that it is a sensation of dryness of the pharynx. The connected more general feeling of lack of water in the body, which is not eliminated by moistening the mouth and throat, is on the other hand not specifically localised.

The feeling of respiratory deficiency (or so-called shortness of breath), which has its own peculiar quality, is reduced by respiratory movements and accordingly localised. Yet the sensations for respiratory obstructions in the lungs are only imperfectly separated from those for circulatory obstructions, if the latter are not combined with palpable alterations in the heart beat. Perhaps this separation is so imperfect only because respiratory disturbances generally evoke an increased heart action as well, and a disturbed heart action makes it difficult to satisfy a respiratory deficiency.

It is moreover to be noted that we have no conception at all, without anatomical and physiological studies, of the form and movements of parts [of the body] having such an extraordinary fine sensitivity, and thereby assured and adept movement, as our soft palate, epiglottis and larynx, since we cannot see them without optical instruments and also cannot easily touch them. We do not indeed yet know, despite all the scientific investigations, how to describe all of their movements with assurance, e.g. not the movements of the larynx which occur in the production of a falsetto voice. Were we innately acquainted with localisations for those organs of ours which are equipped with tactile sensation, we would surely have to expect such an acquaintance just as much for our larynx as for our hands. But in fact our acquaintance with the form, magnitude and movement of our own organs reaches only just as far as we can see and touch them [64].

The extraordinarily varied and finely executed movements of the larynx also teach us something else about the relation between the act of the will and its effect, namely that what we in the first place, and in an immediate manner, understand how to effect is not the innervation of a specific nerve or muscle, nor always a specific position of the mobile parts of our body, but instead the first observable external effect. To the extent that we can ascertain the position of the parts of the body with our eye or hand, this position is the first observable effect with which the conscious intention in the act of the will is concerned. When we cannot do that, as with the larynx and the rear parts of the mouth, the various modifications of the voice, of breathing, of swallowing and so on are these nearest available effects.

Thus the movements of the larynx, although evoked by innervations which are completely alike in kind with those used for moving the limbs, are of no concern in the observation of spatial alterations. One might still ask, however, whether perhaps the very distinct and varied expression of movement which music produces is not reducible to the fact that alteration of pitch in singing is produced by muscle innervations, thus by the same kind of internal activity as is movement of the limbs.

A similar situation also exists with movements of the eyes. We all know very well how to direct our glance to a specific place in the visual field, i.e. how to make the image of that place fall upon the central fovea of the retina. But uneducated people do not know how they move their eyes in doing this, and do not always know how to obey the command of an eye doctor say to turn their eyes to the right, when it is expressed in this form. Indeed, even educated people – although they know how to look at an object held close before the nose and will then squint inwards – do not know how to obey the command to squint inwards without a corresponding object being there.

2. SPACE CAN BE TRANSCENDENTAL WITHOUT THE AXIOMS BEING SO[65]

Nearly all philosophical opponents of metamathematical investigations have treated the two statements as identical, which they by no means are. This has already been explained quite clearly by Benno Erdmann* in the manner of expression current amongst philosophers [66]. I have stressed it myself in an answer directed against the objections of Mr. Land of Leyden**.

Although Albrecht Krause, the author of the latest attack***, cites both of these discussions, with him too the first five of seven sections are still allotted again to defending the transcendental nature of the form of intuition of space, and only two deal with the axioms [67]. This writer is to be sure not merely a Kantian, but an adherent of the most extreme nativist

* *Die Axiome der Geometrie* ['The axioms of geometry'], Leipzig, 1877, ch. III.
** *Mind* 3 (1878), 212.
*** *Kant und Helmholtz*, Lahr, 1878.

theories in physiological optics, and he considers the whole content of these theories to be included in Kant's system of epistemology, for which there is surely not the slightest justification[68], even if Kant's individual opinion – corresponding to the undeveloped state of physiological optics in his time – should have been something of the sort. The question of whether intuition was more or less resoluble into conceptual constructions had at that time not yet been raised[69].

Moreover, Mr. Krause ascribes to me conceptions of local signs, sensory memory, the influence of retinal magnitude, etc. which I have never had and never propounded, or which I have expressly taken pains to refute. As sensory memory[†] I have always characterised only the memory for immediate sensory impressions which are not given a verbal formulation, but I would always have vigorously protested against the assertion that this sensory memory had its seat in the peripheral sense organs. I have performed and described experiments having the purpose of showing that even with falsified retinal images – e.g. when seeing through lenses or through converging, diverging or laterally deflecting prisms – we quickly learn how to overcome the illusion and again see correctly, and then on p. 41 Mr. Krause insinuates to me the view that a child, because his eye is smaller, must see everything smaller than an adult does. Perhaps the present lecture will convince the author named that he has, up to now, completely misunderstood the sense of my empiricist theory of perception.

The objections made by Mr. Krause in his sections on the axioms have in part been disposed of in the present lecture, e.g. the grounds why intuitive representation of an object which has never yet been observed might be *difficult*. He then refers to my assumption, in the lecture on the axioms of geometry[*], of surfacelike beings living on a place or a sphere, which I made in order to render intuitive the relationship between the various geometries; and there follows an explanation that on the sphere two or many 'straightest' [**] lines could indeed exist between two points, but that Euclid's axiom speaks of the one 'straight' line. However, for the surface beings on the sphere the straight line connecting two points on the spherical surface has – according to the assumptions made – no

† ['Sinnengedächtnis']
* See the beginning of this volume.
** I had thus named the shortest or geodetic lines.

real existence at all in their world. The 'straightest' line of their world would be for them precisely what the 'straight' one is for us. Mr. Krause indeed makes an attempt to define a straight line as a line having only one direction. But how should one define direction? – surely only again by means of the straight line. Here we move in a vicious circle. Direction is indeed the less general concept, since every straight line contains two opposed directions.

There then follows an explanation that if the axioms were empirical propositions, we could not be absolutely convinced of their correctness, as surely we are. But this is precisely what the controversy hinges upon. Mr. Krause is convinced that we would not believe measurements testifying against the correctness of the axioms[70]. Here he may well be right as regards a large number of people, who more willingly trust a proposition supported by ancient authority, and closely interwoven with all of their remaining information, than their own reflections. But it should surely be otherwise with a philosopher. People also behaved highly incredulously for long enough towards the spherical shape of the earth, its motion and the existence of meteorites. There is incidentally something correct about his assertion, in that one is well advised to examine all the more rigorously the grounds of proof against propositions of old authority, the longer these propositions have so far proved to be factually correct in the experience of many generations. But the facts must surely decide in the end, and not preconceived opinions or Kant's authority.

It is furthermore correct that if the axioms are laws of nature, they naturally participate in the merely approximate provability of all laws of nature through induction. But the desire to be acquainted with exact laws is not yet a proof that such exist. Odd it is indeed that Mr. Krause rejects the results of scientific measurement because of their limited accuracy, whereas in order to prove, as regards transcendental intuition, that we have no need at all for measurements in convincing ourselves of the correctness of the axioms, he reassures himself with appraisals by visual estimation (p. 62). That is surely measuring friend and foe by different standards! As if any pair of dividers from the worst set of drawing instruments would not yield something more accurate than the best visual estimation, even disregarding the question – which my opponent does not ask himself at all – of whether visual estimation is innate and given *a priori* or whether it is not acquired too.

The expression 'measure of curvature', in its application to three-dimensional space, has given great offence to philosophical writers*. Now this name denotes a certain magnitude defined by Riemann, which when calculated for surfaces coincides with what Gauss called the measure of curvature of surfaces. The name has been retained by geometers as a brief term for the more general case of more than two dimensions. The controversy here concerns only the name, and nothing more than the name, for what is in any case the well-defined concept of a magnitude[71].

3. The applicability of the axioms to the physical world[72]

Here I want to develop the consequences which we would be forced to adopt if Kant's hypothesis of the transcendental origin of the axioms of geometry were correct, and to discuss what value this immediate acquaintance with the axioms would then have for how we judge the relationships in the objective world**.

1. I shall initially stay with the realist hypothesis[73], in this first section, and speak its language, thus I shall assume that the things we objectively perceive really exist and act upon our senses. I do this initially only in order to be able to speak the simple and understandable language of ordinary life and natural science, and thereby to express the sense of my opinion in a manner understandable to non-mathematicians too. I reserve the possibility of dropping the realist hypothesis in the following section, and repeating the corresponding account in abstract language and without any particular presupposition about the nature of the real.

First of all: the sort of alikeness or congruence of spatial magnitudes which might flow, according to the assumption made, from transcen-

* e.g. in A. Krause, *op. cit.*, p. 84.
** So, in order to prevent fresh misunderstandings such as occur in A. Krause, *op. cit.*, p. 84: I am not the one "who is acquainted with a transcendental space having laws proper to itself", but I am instead here seeking to draw the consequences of what I consider to be Kant's unproved and incorrect hypothesis, according to which the axioms are taken to be propositions given by transcendental intuition, and I do this in order to demonstrate that a geometry based upon such intuition would be wholly useless as regards objective knowledge.

dental intuition, must be distinguished from the alikeness in value of such magnitudes which is to be ascertained by measurement with the aid of physical means [74].

I call spatial magnitudes *physically alike in value* [†], if they are ones in which under like conditions, and in like periods of time, like physical processes can exist and run their course [75]. The most frequent process, employed with suitable precautions, for determining spatial magnitudes physically alike in value is the transfer of rigid bodies, as measuring rods or a pair of dividers, from one place to another. It is moreover a quite general outcome of all of our experiences, that if the alikeness in value of two spatial magnitudes has been shown by any adequate method of physical measurement, they show themselves to be alike in value also as regards all other known physical processes [76].

Physical alikeness in value is thus a completely determinate, unambiguous objective property of spatial magnitudes, and obviously nothing hinders us from ascertaining, by experiments and observations, how the physical alikeness in value of a specific pair of spatial magnitudes depends on the physical alikeness in value of other pairs of such magnitudes. This would give us a kind of geometry which, just for the purpose of our present investigation, I will call *physical geometry* [77], in order to distinguish it from the geometry which might be founded upon the hypothetically assumed transcendental intuition of space. Such a physical geometry, implemented in a pure and deliberate manner, would obviously be possible and have completely the character of a natural science.

Even its first steps would lead us to propositions corresponding to the axioms, if we only substitute for the transcendental alikeness of spatial magnitudes their physical alikeness in value.

Namely, as soon as we had found a suitable method for determining whether the separations of any two point pairs were alike (i.e. physically alike in value), we should also be able to distinguish the special case where three points a, b, c lie such that no other point, apart from b, can be found having the same separations from a and c as b does. We say in this case that the three points lie in a straight line.

We would then be in a position to seek three points A, B, C having all three like mutual separation, and which would thus be the corners of an

---

[†] ['physisch gleichwertig' , i.e. physically equivalent; see Translator's Note.]

equilateral triangle. Then we could seek two new points b and c, both having like separation from A, and such that b lies in a straight line with A and B, and c with A and C. Straight away the question would arise: is the new triangle Abc equilateral too as is ABC, thus do we have $bc = Ab = Ac$? *Euclidean* geometry answers: yes; *spherical* geometry asserts that $bc > Ab$ when $Ab < AB$, and *pseudospherical* geometry that $bc < Ab$ under the same condition [78]. Already here the axioms would come up against a factual decision. I have chosen this simple example, because in it we deal only with measuring alikeness or non-alikeness of the separations between points, or as may be with the determinateness or non-determinateness of the situations of certain points, and because no spatial magnitudes of greater composition, straight lines or planes, need to be constructed at all. The example shows that this physical geometry would have its own propositions occupying the place of the axioms.

As far as I can see, even for an adherent of Kant's theory it cannot be a matter of doubt that it would be possible, in the manner described, to give the foundation of a purely empirical geometry, if we did not yet have such a geometry. In it we would deal only with observable empirical facts and their laws. And only inasmuch as the presupposition were correct, that physical alikeness in value always occurs simultaneously for all kinds of physical processes, would the science acquired in such a manner be a theory of space which was independent of the constitution of the physical bodies contained in space.

But Kant's adherents assert that there also exists, besides such a physical geometry, a *pure geometry* founded upon transcendental intuition alone, and that this indeed is the geometry which has been scientifically developed up to now. In this geometry we are said not to deal at all with physical bodies and their behaviour during motions. But we can instead – without knowing anything whatsoever about such [bodies] through experience – form for ourselves, by means of inner intuition, representations of absolutely unalterable and immobile spatial magnitudes, of bodies, surfaces and lines, which even without ever being brought into coincidence through motion (which belongs only to physical bodies) could stand to one another in the relationship of alikeness and congruence*.

---

* Land in *Mind* **2** (1877), 41; A. Krause, *op. cit.*, p. 62.

I will permit myself to emphasise that this inner intuition of the straightness of lines, and alikeness of separations or angles, must have absolute accuracy. Otherwise we would by no means have the right to decide whether two straight lines, when infinitely prolonged, would intersect only once or perhaps even twice (like great circles on a sphere) nor to assert that every straight line which intersects with one of two parallel lines lying in the same plane as itself, must also intersect with the other[79]. One must not try to pass off visual estimation, which is so imperfect, as transcendental intuition, which latter demands absolute accuracy[80].

Suppose the case that we had such a transcendental intuition of spatial structures, and of their alikeness and congruence, and that we could convince ourselves on truly sufficient grounds that we had it. Then a system of geometry would of course be derivable from it, one which would be independent of all properties of physical bodies – a pure, transcendental geometry. This geometry too would have its axioms. But it is clear, even according to Kant's principles, that the propositions of this hypothetical pure geometry need not necessarily concur with those of physical geometry. For the one speaks of the alikeness of spatial magnitudes in inner intuition, the other of physical alikeness in value. This latter obviously depends on empirical properties of natural bodies and not merely on the makeup of our mind.

So one would then have to investigate whether the two mentioned kinds of alikeness necessarily always coincide. This is not to be decided by experience. Has it any sense to ask whether two pairs of dividers encompass, according to transcendental intuition, like or unlike lengths? I do not know how to connect any sense with this; and as far as I have understood recent adherents of Kant, I believe I may take it that they too would answer with a no. Visual estimation, as already said, is something we should not allow to be passed off on us in this respect.

Could one then perhaps infer from propositions of pure geometry that the separations of the tips of the two pairs of dividers are alike in magnitude? For that one would have to be acquainted with geometrical relations between these separations and other spatial magnitudes, and to know directly that these latter were alike in the sense of transcendental intuition. But since one can never know this directly, one also can never obtain it by geometrical inferences.

If one cannot acquire by experience the proposition that the two kinds of spatial alikeness are identical, it must be a metaphysical proposition and correspond to a necessity of thought. But this would then be a necessity of thought which determined not merely the form of items of empirical knowledge, but also their content, as for example with the above construction of two equilateral triangles [81] – a consequence which would indeed contradict Kant's principles. Pure intuition and thought would then yield more than Kant is inclined to concede.

Suppose finally the case that physical geometry had found a series of universal empirical propositions which were identical in formulation with the axioms of pure geometry. It would at most follow from this that it was a permissible hypothesis, leading to no contradiction, that the physical alikeness in value of spatial magnitudes concurs with their alikeness in pure spatial intuition. But it would not be the only possible hypothesis. Physical space and the space of intuition could also be related to each other as is actual space to its image in a convex mirror*.

That physical geometry need not necessarily concur with transcendental geometry follows from our being in fact able to represent them to ourselves as not concurring.

The manner in which such an incongruence would make its appearance results already from what I have explained in an earlier paper**. Let us assume that the physical measurements corresponded to a pseudospherical space. With the observer and the observed objects at rest, the sensory impression of such a space would be the same as if we had before us Beltrami's spherical model in Euclidean space, and the observer were at its centre. But as soon as[†] the observer changed his place, the centre of the sphere of projection would have to migrate with him, and the whole projection be displaced. Thus for an observer whose spatial intuitions and whose estimations of spatial magnitudes had been formed either from transcendental intuition, or as a result of his experience to date in respect of Euclidean geometry, there would arise the impression, that as soon as he himself moved, all objects seen by him were also displaced in a certain manner and variously expanded and contracted in various directions.

---

* See my lecture on the axioms of geometry (at the beginning of this volume).
** on the axioms of geometry (see the beginning of this volume).
† [or perhaps: 'according as'.]

In a similar manner, and in accordance with only quantitatively divergent relationships, we see in our objective world too how the perspective relative situation and the seeming magnitudes of objects at various distances change, as soon as the observer moves. Now we are indeed capable of discerning, from these changing visual images, that the objects surrounding us do not alter their relative mutual situation and magnitudes, as long as the perspective displacements correspond exactly to the law which – according to what has proved true in experience to date – governs them when the objects are at rest; given any deviation from this law, on the other hand, we infer a motion of the objects. In just the same way – as I believe myself permitted to presuppose as an adherent of the empiricist theory of perception – someone who went over from Euclidean to pseudospherical space would also at first indeed believe he saw seeming motions of the objects, but would very quickly learn how to adapt an estimation of the spatial relationships to the new conditions.

But the latter presupposition is one which has been formed only by analogy with what we otherwise know of sense perceptions, and which cannot be tested by making the experiment. So let us assume that the way in which such an observer judged spatial relationships could no longer be altered, because it was connected with innate forms of spatial intuition. Then he would nevertheless quickly ascertain that the motions which he believed he saw were only seeming motions, because of their always being reversed when he himself went back to his starting point; or a second observer could attest that everything remained at rest while the first one changed his place. Thus under scientific investigation – even if perhaps not for unconsidered intuition – it could quickly emerge which spatial relationships were the physically constant ones, somewhat as we ourselves know from scientific investigations that the sum stays fixed and the earth rotates, despite the persistence of the sensory semblance that the earth stays still and the sun goes round it once in 24 hours.

But then the whole of this presupposed transcendental intuition *a priori* would be demoted to the rank of a *sensory delusion*, of an *objectively false semblance*, which we should have to try to free ourselves from and forget, as is the case with the seeming motion of the sun. There would then be a contradiction between what appeared to be spatially alike in value according to innate intuition, and what proved to be such in objective phenomena. The whole of our scientific and practical interest would be attached to the latter. Spatial relationships which were physically alike in value would be portrayed by the transcendental form of intuition only in the way that a flat map portrays the surface of the earth, with very small portions and strips correct but larger ones necessarily false. It would then be not merely a matter of the *manner of appearance*, which of course necessarily occasions a modification of the content to be portrayed, but instead a matter of the relations between appearance and content making the two of them concur on a smaller scale, yet giving a *false semblance* when extended to a larger scale [82].

The conclusion which I draw from these considerations is the following: if there actually were innate in us an irradicable form of intuition of space which included the axioms, we should not be entitled to apply it in an objective and scientific manner to the empirical world until one had ascertained, by observation and experiment, that the parts of space made alike in value by the presupposed transcendental intuition were also physically alike in value. This condition coincides with Riemann's demand that the measure of curvature of the space in which we live should be determined empirically by measurement.

The measurements of this kind performed to date have not yielded any noticeable deviation from zero for the value of this measure of curvature. Thus we can of course regard Euclidean geometry as factually correct within the limits of accuracy of measurement which have been attained up to now [83].

2. The discussion in the first section remained wholly within the domain of the objective and of the realist viewpoint of the enquirer into nature, where the ultimate goal is to apprehend the laws of nature conceptually and where an intuitive acquaintance merely helps to ease matters, or as may be is a false semblance to be done away with.

Now Professor Land believes that in my account I had confused the concepts of the *objective* and of the *real*, that in my claiming that geometrical propositions could be tested and confirmed by experience it was presupposed without foundation (*Mind* 2, 46) "that empirical knowledge is acquired by simple importation or by counterfeit, and not by peculiar operations of the mind, solicited by varied impulses from an unknown reality". Had Professor Land been acquainted with my works on sensations, he would have known that I myself, throughout my life, have combatted against the kind of presupposition which he implies to me. In my paper I did not speak of the distinction between the objective and the real, because it seemed to me that this distinction carried no weight at all in the investigation concerned. In order to give a foundation for this opinion of mine, we will now drop the hypothetical element in the realist view, and demonstrate that the propositions and proofs given so far will even then still have a perfectly correct sense, that one is even then still entitled to ask about the physical alikeness in value of spatial magnitudes and to decide this matter by experience.

The only presupposition we shall retain is that of the law of causality, namely that the representations having the character of perception which come about in us do so according to fixed laws, so that when different perceptions force themselves upon us, we are entitled to infer from this a difference in the real conditions under which they were formed. Concerning these conditions themselves – concerning what is properly real and underlies the appearances – we otherwise know nothing; all opinions which we may otherwise cherish about this are to be regarded as only more or less probable hypotheses. The presupposition we start with, on the other hand, is the basic law of our thought; if we wanted to give it up, we should then renounce altogether the possibility of comprehending these relationships in thought.

I emphasise, that no presuppositions at all about the nature of the conditions under which representations arise should be made here. The hypothesis of subjective idealism would be just as permissible as the realist view, whose language we have used up to now. We could assume that all of our perception was only a dream, although a most highly self-consistent dream in which representation developed out of representation according to fixed laws. In this case the ground for the occurrence of a new seeming perception would have to be sought only in the fact that it had been preceded in the dreamer's mind by the representations of certain other perceptions, and perhaps also by representations of a certain kind of impulses of his own will. That which in terms of the realist hypothesis we call laws of nature, would in terms of the idealist one be laws governing the sequence of successive representations having the character of perception.

Now we find it to be a fact of consciousness, that we believe ourselves to perceive objects which are at specific places in space. That an object should appear at a specific particular place, and not at another one, will have to depend on the kind of real conditions evoking the representation. We must infer that in order to effect the occurrence of the perception of another place for the same object, other real conditions would have had to be present. Thus in the real there must exist some or other relationships, or complexes of relationships, which specify at what place in space an object appears to us. I will call these, to use a brief term for them, *topogenous factors*. Of their nature we know nothing; we know only that the coming about of spatially different perceptions presupposes a difference in the topogenous factors.

Besides this there must, in the domain of the real, be other causes which effect our believing that at the same place we perceive at different times different material things having different properties. I will allow myself to give them the name of *hylogenous factors*. I chose these new names in order to exclude the involvement of any connotations which might be attached to customary words [84].

Now when we perceive and assert anything which states a mutual dependence of spatial magnitudes, then without doubt the factual sense of such a statement is merely that between certain topogenous factors, whose proper nature however remains unknown to us, there exists a certain lawlike connexion, whose type is likewise unknown. For precisely this reason, Schopenhauer and many adherents of Kant have come to the incorrect conclusion that there is altogether no real content in our perceptions of spatial relationships – that space and its relationships are only a transcendental semblance without there corresponding anything actual to them [85]. But we are at any rate entitled to apply the same considerations to our spatial perceptions as to other sensory signs, e.g. to colours. Blue is only a mode of sensation; but that we should see blue in a specific direction at a certain time is something which must have a real ground. Should we at some time see red there, this real ground will have to have altered.

When we observe that physical processes of various kinds can run their course in congruent spaces during like periods of time, this means that in the domain of the real, like aggregates and sequences of certain hylogenous factors can come about and run their course in combination with certain specific groups of different topogenous factors, such namely as give us the perception of parts of space physically alike in value. And when experience then instructs us, that any combination or sequence of hylogenous factors which can exist or run its course in combination with one group of topogenous factors, is also possible with any other physically equivalent[†] group of topogenous factors – then this is at any rate a proposition having a real content, and thus topogenous factors undoubtedly influence the course of real processes.

In the example given above of the two equilateral triangles, it is only a matter of (1) the alikeness or non-alikeness, i.e. physical alikeness or non-alikeness in value, of distances between points; (2) the determinateness or non-determinateness of the topogenous factors of certain points. But these concepts of determinateness and of alikeness in value in respect of certain sequences can also be applied to objects whose nature is otherwise quite unknown. I infer from this that the science I have called physical geometry contains propositions having real content, and that its axioms are determined not by mere forms of representation, but by relationships in the real world[86].

This still does not entitle us to declare impossible the assumption of a geometry founded upon transcendental intuition. One could e.g. assume that an intuition of the alikeness of two spatial magnitudes might be produced immediately, without physical measurement, by the influence of topogenous factors on our consciousness, thus that certain aggregates of topogenous factors might even be equivalent in respect of a psychic, immediately perceivable effect. The whole of Euclidean geometry is derivable from the formula giving the distance between two points as a function of their rectangular coordinates. Let us assume that the intensity of that psychic effect, whose alikeness appears in our representation as alikeness of distance between two points, depends in the same way on some or another three functions of the topogenous factors of any point, as does the distance in Euclidean space on the three coordinates of any point. Then the system of pure geometry of such a consciousness would have to satisfy the axioms of Euclid, however else the topogenous factors of the real world and their physical equivalence behaved.

It is clear, that in this case one could not decide from the form of intuition alone whether there was concurrence between psychic and physical alikeness in value for spatial magnitudes. And if it should turn out

---

† ['äquivalent'].

that there was concurrence, the latter would have to be regarded as a law of nature, or (as I have termed it in my popular lecture) as a pre-established harmony between the world of representations and the real world – just as much as it rests upon laws of nature that the straight line described by a light ray coincides with the one formed by a taut thread.

With this, I consider myself to have shown that the proof which I gave in Section 1.1 in the language of the realist hypothesis reveals itself to be valid even without the latter's presuppositions.

When we want to apply geometry to facts of experience, where it is always only a matter of physical alikeness in value, we can apply only the propositions of that science which I have termed physical geometry. For anyone who derives the axioms from experience, our geometry up to now has indeed been a physical geometry, only one which relies on a great host of experiences collected without any plan, instead of on a system of methodically pursued ones. One should moreover mention that this was already the view of Newton, who declares in the introduction to his *Principia*: "Geometry itself has its foundation in mechanical practice, and is in fact nothing other than that part of the whole of mechanics which accurately states and founds the art of measurement."*

On the other hand, the assumption that there is an acquaintance with the axioms which comes from transcendental intuition is:

(1) an *unproved* hypothesis;

(2) an *unnecessary* hypothesis, since it does not pretend to explain anything in what in fact is our world of representations which could not also be explained without its help;

(3) a *wholly unusable* hypothesis for explaining our acquaintance with the actual world, since the propositions laid down by it should only ever be applied to the relationships in the actual world after their objective validity has been experimentally tested and ascertained.

Kant's doctrine of the *a priori* given forms of intuition is a very fortunate and clear expression of the state of affairs; but these forms must be devoid of content and free to an extent sufficient for absorbing any content whatsoever that can enter the relevant form of perception.

---

* Fundatur igitur Geometria in praxi Mechanica, et nihil aliud est quam Mechanicae universalis pars illa, quae artem mensurandi accurate proponit ac demonstrat. [Somewhat more literally than in Helmholtz' version: "Geometry is therefore founded... part of universal mechanics... sets out and demonstrates...".]

But the axioms of geometry limit the form of intuition of space in such a way that it can no longer absorb every thinkable content, if geometry is at all supposed to be applicable to the actual world. If we drop them, the doctrine of the transcendentality of the form of intuition of space is without any taint[87]. Here Kant was not critical enough in his critique; but this is admittedly a matter of theses coming from mathematics, and this bit of critical work had to be dealt with by the mathematicians.

NOTES AND COMMENTS

[1] The rector's address on 'The Facts in Perception' , given on the occasion of the anniversary celebrations of Berlin University in 1878, rightly counts as the richest in content amongst Helmholtz' epistemological studies. Although Koenigsberger's comment (*H. v. Helmholtz*, vol. II, p. 78) that Helmholtz here gave "a self-contained system of his philosophy" seems to go a little too far, it is certain enough that the lecture contains the most complete and rounded off presentation of his epistemology. Form and content are on a level such that one can willingly concur with Koenigsberger's declaring the study to be the "finest and most important" of his adresses (*ibid.*, p. 246).

How great a fundamental importance Helmholtz himself ascribed to the subject of the address is perhaps best discernible from the sonorous titles which he passingly considered for introducing the lecture. "I shall devise the title last of all," he writes to his wife, "I do not yet know it. Perhaps 'What actually exists?' or 'Everything Perishable is only a Likeness' or 'A Journey to the Mothers'." (*ibid.*, p. 246)

[2] The predicate 'true' should be ascribed in the strictest sense only to judgements (statements). One can thus ask about the truth of *thought*, to the extent that 'thinking' is intended to mean the same as 'judging' . But if we talk about what is true in our *intuition*, this manner of speech refers only to judgements *about* our intuition, or to judgements – should there be such – which intuition contains in a concealed manner.

[3] This question can be included under the preceding one, since the importance for scientific enquiry and for life of the relationship between our intuitions and actuality is precisely that one must be acquainted with this relationship in order to pass true judgements upon actuality.

[4] Helmholtz' conceptions seem to presuppose that one somehow defines philosophy as the science of the *mental*. Yet such a definition would hardly be suitable, since the mental – here obviously thought identical with the 'psychic' – indeed forms the object of a particular science, namely psychology. Of course, a simple identification of psychological and philosophical enquiry has occasionally been attempted (thus Theodor Lipps wanted to define philosophy as "mental science[†] or the science of inner experience"). But in this way

---

[†] ['Geisteswissenschaft' : the distinction at English-speaking universities between 'arts' and 'sciences' is spoken of in German as a distinction between 'mental sciences' and 'natural sciences'. Lipps obviously has the literal sense of this term in mind, and M.S. alludes to it in his comments.]

one assuredly neither does justice to philosophy nor to psychology: many of course, and not without justification, consider psychology to be a branch of natural science. Without here going into the controversial and basically unessential issue of the definition of philosophy, we can say that the wholly correct thought that Helmholtz here has in mind would be better formulated as follows: the natural sciences and the mental sciences, when they pursue their proper issues down to ultimate principles, necessarily end up in epistemology and meet there, thus in the domain of philosophy.

One also has to reach this domain to decide whether, or in what sense, the customary distinction between the mental and the corporeal can be maintained. When Helmholtz contrasts here the "influences of the corporeal world" and the "mind's own workings", he thereby attaches himself verbally to the popular conception of the world. Whether or not he thereby perhaps also uncritically introduces presuppositions having a bearing on the matter, will only emerge in the further course of the discussion (compare below note 49).

[5] One must be carefully on one's guard against misunderstanding the meaning of the term 'prior to experience' in Kant's philosophy. The word 'prior' has namely two meanings, since it may be understood temporally or logically. For Kant, a priori knowledge is only in the *logical* sense *prior* to experience, i.e. its validity is not dependent upon experiences, does not have its logical ground in them. Kant says emphatically (*Kritik der reinen Vernunft* ['Critique of pure reason'], 2nd ed., p. 1): "Thus in respect of time no knowledge comes about in us prior to experience, and any such begins with the latter. But though all of our knowledge starts *with* experience, it yet does not therefore simply all arise *out of* experience." [†]

[6] In place of 'transcendental' one should properly have '*a priori*' . It has already been mentioned (note I.55) that Helmholtz often exchanged the two terms. For Kant *a priori* is whatever is valid independent of experience. The word transcendental has a somewhat more complicated meaning. Kant's own explanation of it (*Kritik der reinen Vernunft* [*op. cit.*], 2nd ed., p. 25) runs: "All knowledge which deals not so much with objects, as rather with our manner of knowing objects, insofar as this is taken to be one possible *a priori*, I call transcendental." And at another point (*ibid.*, p. 81) he says expressly: "Therefore neither space, nor any geometrical specification of it *a priori*, is a transcendental representation; but rather only the knowledge that these representations are not at all of empirical origin, and the possibility whereby it can[††] nonetheless relate *a priori* to objects of experience, can be called transcendental."

[7] According to Locke (1632–1704) there are two[†††] kinds of properties. Namely the quantitatively conceivable spatial and temporal ones, which in his view belong to things themselves and are termed by him 'primary' properties, and the 'secondary qualities' such as red, cold, loud and so on. These differ from the former in being only modifications of our perceiving consciousness, in which they are produced by the influence of bodies upon our sense organs. They are thus not properties of things at all, but belong only to the subject and have no similarity to the true properties of external things. This is the doctrine of the 'subjectivity of the qualities of the senses', which was already quite clearly expressed

---

[†] [Kemp Smith's translation is not followed exactly.]
[††] [Some read 'können' for 'könne', i.e. 'they can' instead of 'it can'.]
[†††] [In fact three kinds: see remark appended to note 24 below.]

in antiquity by Democritus. In recent times it was already again advocated before Locke, and probably first of all by Galileo.

[8] What Johannes Müller termed the law of specific sensory energies was admittedly not formulated by him in a wholly unobjectionable manner. One can perhaps express it more carefully thus: the modality of the sensation depends in an immediate manner only upon what region of the central organ is put into a corresponding excited state, independent of the external causes bringing about the excitation. Probably still better and more general formulations are possible – but what is properly the basic thought, one expressing nothing other than the doctrine of the subjectivity of the qualities of the senses in a physiological formulation, remains the same and remains correct. The attacks *in principle* which have been levelled against J. Müller's law, e.g. by W. Wundt in his psychology, are therefore in any case unjustified.

[9] One no longer today, according to the electromagnetic theory of light (to whose victory precisely Helmholtz himself contributed), conceives of light oscillations as material motions, and they are consequently regarded as essentially different from sound waves. But this fact is of no significance for the comparison drawn by Helmholtz here.

[10] According to the theory of Young and Helmholtz, there are three different basic *processes* in the retina of the eye, which correspond to the sensations red, green and blue. The sensations of the remaining colours occur, according to this theory, when more than one of these processes are excited at once. But psychologically speaking, every colour sensation is doubtless something simple, unanalysable. One should thus strictly only talk of a mixture of the physiological processes, not of a mixture of the sensations or 'in sensation' . We shall find on a number of further occasions that Helmholtz did not distinguish strictly enough between the physiological process of sensory excitation and the psychological process of sensation.

It may be noted incidentally that modern theories of colour (Hering, v. Kries, G. E. Müller), in order to achieve a better fit with the empirical facts, have deviated not inconsiderably from the view of Young and Helmholtz. Yet that is wholly inessential as regards the epistemological connexions.

[11] Helmholtz' theory of consonance [harmony], to which he refers in this paragraph, has not remained unchallenged. Stumpf in particular, in his *Tonpsychologie* ['Psychology of sound'], has developed a view based on essentially different foundations. But once again: in this connexion the correctness of the psycho-physiological theory is not involved at all, as it only has to fulfil the task of an explanatory example.

[12] The formulation of this sentence could be attacked from various aspects. Here we shall only point out that terming sensations effects 'in our organs' is dubious, since – as psychic magnitudes – they are obviously not in our organs so much as in our *consciousness*. We have before us the substitution of one thing for another which was already touched upon a moment ago in note 10, and one which would remain harmless only if one could demonstrate that Helmholtz always understood the word sensation[†] only 'in a physical respect',

---

[†] ['Empfindung' : this word has no etymological connexion with the German equivalents of 'sense' (noun) and 'sense organ' , although it generally corresponds to 'sensation' and the verb 'empfinden' to 'to sense' .]

as a process in the sense organ. (Compare B. Erdmann, *Die philosophischen Grundlagen von Helmholtz' Wahrnehmungstheorie* ['The philosophical foundations of Helmholtz' theory of perception'], p. 19.) Where he does this, he contrasts with it the corresponding psychic datum as a 'perception'. It is put thus in an early note made known by Koenigsberger (*Hermann von Helmholtz*, vol. II, p. 129): "Perception is the becoming conscious of a certain sensation, i.e. of a certain state of our organs." Yet he seems all the same not to adhere consistently to this terminology and manner of thought.

¹³ It is not only the expression of the sensation, but the latter itself which is different; thus instead of "how such an effect expresses itself" one should indeed rather say "what sort of effect occurs".

¹⁴ In accordance with note 2 above, talk about truth of *images* is to be understood as an abbreviated manner of speech whose sense is: truth of judgements which assert alikeness of objects with images. Elsewhere (1st ed. of the *Physiologische Optik* [*op. cit.*], reprinted 3rd ed., vol. III, p. 18) Helmholtz analogously explains true representations to be such as lead us to suitable† conduct; successful conduct is namely an indication of the correctness of the judgements which are taken as a basis: "It is therefore my opinion, that there can be no possible sense at all in speaking of any other kind of truth of our representations than a *practical* one. Our representations of things *cannot* be anything at all other than symbols – naturally given signs for the things which we learn to use for regulating our movements and conduct. If we have learnt to read those symbols correctly, we are in a position to arrange our conduct with their help such that it will have the desired outcome, i.e. such that the expected new sensations occur. Not only is there *in actuality* no other manner of likening representations and things, as all schools are agreed, but none other is even *thinkable* at all and has at all any sense."

¹⁵ In Schlick, *Allgemeine Erkenntnislehre* ['General epistemology'], part I, an attempt is made to show that forming such an image of what is lawlike in the actual, with the help of a sign system, altogether constitutes the essence of all knowledge, and that therefore our cognitive process†† can only in this way fulfil its task and needs no other method for doing so.

¹⁶ The way in which this sentence is formulated, together with what follows, creates the impression that Kant declared the qualities of sensation to be a 'form of intuition'. That is not the case by any means. Only space and time are forms according to Kant, while the qualities of sensation are for him always *contents* of intuition, and have a wholly different significance for cognition from what the forms have (only the latter are namely for him sources of synthetic judgements *a priori*). Helmholtz' remark is simply intended to signify that the qualities of sensation are purely *subjective*.

---

† ['zweckmässig': the word mystifyingly rendered 'purposive' in translations of Kant's *Critique of Judgement*.
   In the quotation which follows, Helmholtz appears to make practical success not a mere criterion of the truth of something, but instead the *meaning* of its being true. However, the questions which introduced the discussion in the present paper ("What is true in our intuition and thought? In what sense do our representations correspond to actuality?") indicate that here he takes *correspondence* to be the meaning of truth, even if practical success is the only criterion for the existence of this correspondence.]
†† ['unser Erkennen']

[17] This sentence need not be interpreted to mean that Helmholtz maintained the existence of so-called 'innervation sensations', which is mostly denied by modern psychology. He refers only to the clearly indubitable fact, that in executing a voluntary movement we indeed have a consciousness of initiating that movement. We moreover become conscious of alterations in situation and of movements of our limbs themselves by way of the so-called kinaesthetic sensations, which are presumed to be transmitted by special sensory nerves ending in the muscles, sinews and joints.

[18] It is obvious that here Helmholtz even expressly contests – by implication – the existence of special innervation sensations, since his opinion of course is that consciousness of the impulses is not characterised by certain specific qualities of sensation, but by our representations of the movements connected with them. Anyway, the perhaps detailed psychological assumptions made here by Helmholtz are, as also B. Erdmann considers (*op. cit.*, p. 26), "of no significance for the essential content of his theory".

[19] The sense of this and the preceding sentence is probably understandable, but the formulation is again impaired by the tendency to blur the distinction between the sensation as a psychic datum and as an excitation of the sense organ. The contrast made by Helmholtz between the "perceptions of *psychic* activities" and the "sensations of the outer senses" seems only to be possible if the latter are taken to be non-psychic. But this is contradicted by the next two sentences, in which he obviously talks of qualities of sensation as being contents of consciousness. Had Helmholtz contrasted the 'outer' senses with 'inner' (instead of 'psychic') activities, this would have corresponded to the distinction laid down by him above between outer and inner intuition. But this distinction too is dubious on various grounds (compare Schlick, *Allgemeine Erkenntnislehre* [*op. cit.*], § 19). The only legitimate way of doing justice to the distinction developed here by Helmholtz, between spatial and non-spatial experiences, probably consists in classing as the former *all sense perceptions* whatsoever, and as the latter simply all remaining contents of consciousness.

The statement that sensations "proceed subject to some kind of innervation or another" can be elucidated by examples: tactual perceptions generally require a movement of the touching hand, visual perceptions a movement of the seeing eye, of the head, etc. With perceptions of sounds, where this holds to a lesser extent (although even here bending one's ear or approaching towards the source of sound, and so on, play a part), the spatial specificity is also straight away less pronounced.

[20] Here again as above, the expression 'form of intuition' is used in a wholly different sense from that in Kant (compare note 16). But Helmholtz is quite right in coordinating the spatial intuition described by him with the qualities of sensation, since in both cases it is a matter of subjective, psychic contents. What he namely describes, in what precedes and what follows, is *psychological space* (or properly the psychological *spaces* – since one must separate the spatial data of e.g. sensations of movement from those of visual or of tactual perceptions as something wholly different, even though they are all connected by close associations) and not physico-geometrical space. The latter is a non-qualitative, formal conceptual construction; the former, as something intuitively given, is in Helmholtz' words imbued with the qualities of the sensations, and as purely subjective as these are.

[21] In fact, although e.g. the sweet taste of sugar is purely subjective, it is anything but a semblance – as a quality of sensation it is on the contrary something of the most indubi-

table actuality. Likewise the spatial properties of perceptions, with whose help we indeed orientate ourselves in our surroundings with the greatest assurance. A semblance† would be present if our perceptions seduced us into mistaken behaviour towards things, as say when we grasp for the mirror image of an object because we falsely take it for the object itself. Compare note 31.

²² According to this wholly correct explanation, the judgement that outer intuition is spatial is *necessarily* valid, because it is analytic (compare note I.2). For we *term* "outside" precisely that which has some spatial specification in the described sense. Naturally, our own body belongs here too, as it is indeed conceived of as something external by contrast with self-consciousness.

²³ The manner of being *a priori* which Helmholtz ascribes, in the words of the text, to space must be termed a 'psychological' one. It is not unimportant to ascertain whether, or to what extent, his conception coincides with Kant's doctrine, to which Helmholtz expressly makes reference and from which he starts.

Now there are two different contrasting interpretations of Kant's apriorism. Firstly, the one chiefly taken into account by Helmholtz himself (and also advocated e.g. by Schopenhauer), namely the psychological conception of it, which considers the most essential feature of *a priori* cognitions to be their being conditioned by the psychic makeup of the cognising consciousness. Secondly, the transcendental-logical exegesis, according to which the essence of the *a priori* consists in its comprising the ultimate axioms which alone form the foundation for all rigorous cognition and guarantee the latter's validity. An unbiased reading of Kant's writings seems to teach us that the second point of view is the more important for his system, while yet appearing as something closely and not quite dissolubly interwoven with the first point of view. Thus neither does the one-sidedly psychological conception do justice to Kant's thought, nor either the logical interpretation of the so-called 'neo-Kantian' schools, which want rigorously to exclude everything psychological from any connexion with the *a priori*.

In Helmholtz' account, as already said, spatial intuition is acknowledged to be *a priori* only in the psychological sense, so that his epistemology thoroughly deviates from Kant's, as he himself also plainly perceived. B. Erdmann too, in his last work (*op. cit.* in note 12, p. 27), judges that space is according to Helmholtz a subjective form of intuition "in a sense thoroughly foreign to Kant's doctrine of space", and he justly claims that here "Kant's rationalist thoughts" are "twisted round into their empiricist counterpart".

²⁴ A considerable part of the address on 'The Facts in Perception' was repeated by Helmholtz word for word in §26 of the second edition of his *Physiologische Optik* [*op. cit.*]. There (p. 588) the present passage in followed by some statements taken over from the first edition, for which room may be found here on account of their epistemological interest:

"As regards, in the first instance, the *properties* of the objects of the external world, a little reflexion shows that all properties ascribed to them by us only characterise *effects* which they exert either upon our senses or upon other objects in nature. Colour, sound, taste, smell, temperature, smoothness and solidity belong to the first class, they characterise effects

---

† ['Schein': this word is rendered 'illusion' by Kemp Smith in his translation of the *Critique of Pure Reason*, and similarly understood here by M.S. But Helmholtz does not seem to understand by it something necessarily illusory.]

upon our sense organs. Smoothness and solidity characterise the degree of resistance which bodies we contact offer either to a sliding contact or to the pressure of the hand. However, other natural bodies too can take the place of the hand, likewise for examining other mechanical properties, elasticity and weight. Chemical properties as well are related to reactions, i.e. effects which the natural bodies under consideration exert upon others. It is likewise with the other physical properties of bodies – optical, electrical, magnetic.

"Everywhere we deal with mutual relations between one body and another, with effects upon one another which depend on the forces which different bodies exert upon one another. For all natural forces are forces which one body exerts upon the others. If we think of mere matter without forces, it is also without properties – apart from its varied distribution in space and its motion. For that reason too, no property of natural bodies makes its appearance until we involve them in the corresponding mutual effect with other natural bodies or with our own sense organs. But since such a mutual effect can occur at any moment, or as may be can also be brought about at any moment by our will, and we then always see the appropriate kind of mutual effect occur, we ascribe to the objects a lasting capacity for such effects, and one which is always ready to become effective. This lasting capacity we call a *property*[†].

"Now it results from this that the *properties* of natural objects, despite this name, in truth characterise nothing whatsoever proper to the individual object in and for itself, but instead always a relation to a second object (which includes our sense organs). The kind of effect must naturally always depend on de peculiarities both of the body exerting an effect and of the body upon which an effect is exerted.

"We do not have even a moment's doubt about this when speaking about properties of bodies such as emerge if we have two bodies, belonging both to the external world, and one exerts an effect upon the other, e.g. in chemical reactions. Concerning the properties, on the other hand, which are based upon mutual relations between things and our sense organs, people have always tended to forget that here too we have to do with a reaction against a particular reagent, namely our nervous apparatus, and that colour, smell and taste, the feelings of warmth and of cold are also effects which depend quite essentially upon the kind of organ on which an effect is exerted. Of course, the reactions of our senses to natural objects are the most frequently and universally perceived; they have the most predominant importance for our well-being and comfort; the reagent against

---

[†] [Helmholtz in effect distinguishes three kinds of attribute that bodies may have. There are firstly intrinsic attributes of matter, namely its distribution and motion. These are opposed to properties, as just defined, and the latter are distinguished into those involving a mutual effect between bodies one of which is a sense organ, and those where a sense organ is not one of the bodies concerned.

This recalls Locke's distinction between primary, secondary and tertiary qualities (*Essay concerning Human Understanding*, book II, ch. 8). The first are "such as are utterly inseparable" from bodies (§9), namely the "bulk, figure, number, situation, and motion or rest of their solid parts" (§23). Secondary qualities of bodies are "powers to produce various sensations in us... by the bulk, figure, texture, and motion of their insensible parts, as colours, sounds, tastes, etc." (§10). The third kind are powers in a body to change "the bulk, figure, texture, and motion of *another* body, as to make it operate on our senses differently from what it did before" (§23). Locke also complains (as Helmholtz will below) that the similarity between the second and third kinds is commonly overlooked.

A clear difference between their views is that Helmholtz makes fixity (solidity) entirely an attribute of the second kind.]

which we have to try them out is endowed upon us by nature – but this does not change the situation.

"It is for this reason senseless to ask the question of whether cinnabar actually is red as we see it, or whether this is only an illusion of the senses. The sensation of red is the normal reaction of normally formed eyes for light reflected by cinnabar. Someone colour-blind to red will see the cinnabar as black or dark greyish yellow, and this too is the correct reaction for his particular kind of eye. He must only know that his eye is indeed of a different kind from those of other people. The one sensation is not in itself any more correct or false than the other, even if those seeing red have a great majority on their side. In general, the red colour of cinnabar only exists inasmuch as there are eyes constituted in a way similar to those of the majority of people. With equal justice, it is a property of cinnabar to be black, namely for people colour-blind to red. In general, light reflected by cinnabar is not in itself to be called red at all, it is only red for certain kinds of eyes.

"When speaking of properties of bodies which these have in respect of other bodies in the external world, we do not forget to characterise that body too in respect of which the property occurs. We say: 'Lead is soluble in nitric acid, it is not soluble in sulphuric acid.' Should we merely want to say 'Lead is soluble', we would immediately notice that this is an incomplete assertion, and would immediately have to ask in what it is soluble. But when we say 'Cinnabar is red', it is automatically understood implicitly that for our eyes it is red, and for the eyes of other people, which we presuppose to be constituted alike. We believe we need not mention this, and for that reason may well also forget it, and may be mislead into believing that redness is a property belonging to cinnabar, or to light reflected from it, quite independent of our sense organs.

"It is something else if we assert that the wavelengths of light reflected from cinnabar have a certain length. That is a statement which we can make independent of the particular nature of our eye. With this statement, however, it is then also only a matter of relations between the substance and the various aether-wave systems."

[25] In a quite similar manner, H. Poincaré (*Der Wert der Wissenschaft* ['The value of science'], 2nd ed., pp. 61 f.) elucidates what is peculiar to spatial alterations (movements), as opposed to a qualitative change (e.g. alteration of a body's colour) in our surroundings.

[26] See e.g. Fichte, *Grundlage der gesamten Wissenschaftslehre* ['The foundation of the whole theory of science'], §2, no. 9.

[27] Helmholtz speaks, as is customary, simply of 'spatial intuition', although – as already remarked (note 20) – one must properly distinguish as many spatial intuitions as there are senses: e.g. the blind man precisely *lacks* intuitive visual space. But there occurs a very close associative connexion between the spatial intuitions, and it may be asked whether one specific sensory domain plays an outstanding role in this, so that it as it were supplies the nucleus of association about which the remaining spatial representations group themselves. One should no doubt answer affirmatively, and it is in the first place – as is also clear from Helmholtz' own account on pp. 123 f. – the data of sensations of movement to which that central position must be assigned. We may understand Poincaré too in this sense, when he says (*Der Wert der Wissenschaft* [*op. cit.*], p. 71: "Actual space is the space of movement." In the second place, as Helmholtz emphasises, one should of course above all take the sense of touch into account, whose representations are moreover most closely connected with those of the sense of movement.

[28] A continuum of n dimensions is characterised by the fact that it can be completely

split by a continuum of $(n-1)$ dimensions. Poincaré too (*Wissenschaft und Hypothese* ['Science and hypothesis'], pp. 33 f.; *Der Wert der Wissenschaft* [*op. cit.*], p. 73) uses this feature as a characteristic of the *n*-dimensional continuum.

[29] This relative clause shows how much the terms taken over from Kant's philosophy have changed their sense in Helmholtz. According to Kant, all empirical objects are naturally without exception "tied to the form of the human faculty of representation", since they of course only attain objectivity through precisely this form. Helmholtz here obviously wants to indicate that the spatial properties of bodies indeed have a certain objective significance over and above the "faculty of representation". This would perhaps correspond to Kant's opinion insofar as he too held the view that there must for every spatial specification "be also in the object, which is in itself unknown, a ground" (*Metaphysische Anfangsgründe der Naturwissenschaft* ['Metaphysical principles of natural science'], Dynamik, theorem 4, note 2; compare also A. Riehl, *Der philosophische Kritizismus* ['Philosophical criticism'], 2nd ed., vol I, p. 470).

[30] The formulation of the text, as literally understood, is highly open to attack. It would necessarily lead to misunderstandings and confusions, on grounds already raised more than once (compare note 12), were it not mitigated by the following sentence (see the next note). The *qualities* of sensation certainly belong as such to our consciousness alone, and in no way to the nervous system. If I sense a bitter taste or hear a loud note, my nerves are not then bitter or loud.

Helmholtz expresses himself as if our consciousness with its sensations were located in the nervous system, and consequently also in space – since the nervous system is of course an object in space. One would thus arrive at a 'projection theory' of perception, according to which the qualities of sensation are firstly sensed in the body (nervous system) itself, so as thereupon to be "projected out" into space. (This impossible philosophical projection theory should not be confused with a certain purely physiological theory of spatial vision, which – if indeed unsuitably – has likewise been given the name projection theory. Compare on this e.g. von Kries in vol. III of the third edition of Helmholtz' *Physiologische Optik* [*op. cit.*], p. 466.) As the spatial (according to Helmholtz himself too) is a form of intuition of our consciousness, it is not proper then to localise our consciousness in turn somewhere in intuitive space.

If one nevertheless attempts this, one gets involved in insoluble contradictions and commits the error which R. Avenarius characterised as 'introjection', and which he has illuminated with great acuity – and shown to be avoidable – in his writings *Kritik der reinen Erfahrung* ['Critique of pure experience'] and *Der menschliche Weltbegriff* ['The human concept of the world']. Compare on this Schlick, 'Idealität des Raumes, Introjektion und psychophysisches Problem' ['The ideality of space, introjection and the psychophysical problem'], *Vierteljahresschr. für wiss. Philosophie* **40**.

[31] This sentence perhaps does not wholly exclude all misunderstanding. But it still emphasises with welcome plainness that what Helmholtz has just characterised as a semblance – namely that the objects extant in space are imbued with the qualities of sensation – that this is in fact 'the original truth' , because precisely these qualities actually offer themselves in intuitive space and have an existence nowhere else.

If one understands by 'bodies' the objects perceived in intuitive space, we must bear in mind with Kant that they "are not something outside us, but merely representations in us, and hence that it is not the motion of matter that effects representations in us, but that this motion is instead itself a mere representation" (*Kritik der reinen Vernunft* [*op. cit.*], 1st ed., p. 387). And in the same sense E. Mach (*Analyse der Empfindungen* ['The analysis of sensations'], 5th ed., p. 23) says: "It is not bodies which generate sensations, but it is sensation complexes instead which form bodies." Thus the 'external causes' of which Helmholtz, on p. 121, considered sensations to be effects, are at any rate not these bodies, but can instead only be understood to be transcendent[†] things (although the concept of a cause cannot, according to Kant, be applied also to the latter). That these things are not subjects of qualities of the senses is something which Helmholtz obviously wanted to stress in the preceding sentence.

It remains somewhat unsatisfying that Helmholtz does not sharply emphasise that there is in *no* way a 'semblance' when we perceive sensory qualities as localised in intuitive space.

[32] We may ask ourselves, on looking back through the last few pages of Helmholtz' account, what constitutes these "most essential features" of spatial intuition. In which case we are probably led to the statement that space is a three-dimensional continuous manifold, in which there is an enduring existence of different things at the same time one beside another, and in which magnitudes can be likened with one another.

[33] Helmholtz here introduces a distinction of the greatest importance for his theory, namely that between the 'general form of spatial intuition' and its 'narrower specifications' , which latter are expressed in the axioms of geometry. For him, the general form is that 'schema devoid of any content' which he declared on pp. 1–2 to be the true form of intuition, in respect of which Kant's doctrine of the *a priori* is to be upheld.

One must ask: what then are the *broader* specifications, opposed to these 'narrower' ones, by which the 'general form' is supposed to be characterised? For some or other characteristics must surely belong to it too, since one could not otherwise at all speak of it as of something specific. To this question there seem to be two possible answers. The characteristics sought for might firstly consist of certain peculiarities of sense perceptions which could not be further described but only displayed and witnessed, ones which precisely endow these with the character of spatiality – say the 'extendedness' of a visual perception, or the wholly different 'extendedness' of a tactual perception. While secondly, one might be supposed to look for the demanded specifications of the 'general' form of spatial intuition in precisely those 'most essential features' of which Helmholtz has already spoken, and which were assembled in the preceding note.

In the second case[††], however, there arises the question of whether those general features could and should, just as well as the 'narrower specifications' , be formulated in certain geometrical axioms. Modern geometry is inclined to answer this question affirmatively, and one would then be unable to uphold, in the required sense, the distinction between general and particular specifications of spatial intuition. Thus to make this possible, the

---

[†] [This term is not a synonym of 'transcendental' (see note 6), but used by Kant for "a principle professing to go beyond the limits of possible experience", see *Kritik der reinen Vernunft* [*op. cit.*], 2nd ed., p. 352. By a 'transcendent thing' M.S. means an object whose properties might thus purportedly be characterised.]

[††] [which seems to be what Helmholtz intends, since the whole of the previous discussion is supposed to establish the limit up to which "the approach of natural science can take the same path" as Kant.]

'general form' will have to be understood as that indescribable psychological component of spatiality which imbues sense perceptions.

Poincaré (*Der Wert der Wissenschaft* [*op. cit.*], p. 48) raises the question of whether perhaps space is "a form forced upon our consciousness" as far as its purely *qualitative* specifications are concerned. The properties of this 'general form' or this 'schema devoid of any content' would then have to be expressed in the propositions of *analysis situs* (see above, note I.21). But Poincaré arrives at the result that these propositions too may be presumed to rest upon experiences.

Some neo-Kantians (as P. Natorp, E. Cassirer) have tried to conceive of the *a priori* nature of spatial intuition in the genuine sense of Kant (thus not in Helmholtz' psychological interpretation), but such that it does not comprise the stipulation of some or other specific Euclidean or non-Euclidean geometry. They seem, however, to be defeated by their effective failure to say what are the *a priori* laws of spatial intuition which, in their opinion, then still remain. Compare Schlick, 'Kritizistische oder empiristische Deutung der neuen Physik?' ['A criticist or an empiricist interpretation of the new physics?'], *Kantstudien* **26**.

³⁴ The following basic propositions, however, do not yet form a complete system of axioms upon which the whole of geometry could be built without the assistance of further propositions. See note I.6. Compare with the subsequent account in the text the lecture on 'The Axioms of Geometry' .

³⁵ Here and in what follows one might more correctly, as before, put '*a priori*' in place of 'transcendental' .

³⁶ Since Helmholtz speaks here of the form of intuition of the *eye*, it follows that in his opinion too each individual sense has basically its own particular form of intuition, in the significance in which he uses this phrase. It is of interest here, considering what was said in note 33, that von Kries (in the third edition of Helmholtz' *Physiologische Optik* [*op. cit.*], vol. III, p. 499) has expounded the view that the spatial intuition of the eye rests in all of its *quantitative* specifications upon experience, while on the other hand its *qualitative* situational properties (thus the data to be dealt with by analysis situs) are given in a fixed manner in advance by physiological laws of formation: "Thus what we can think of as made fixed by laws of formation would still leave alterable the arrangement in the visual field, in a manner similar to that in which a picture painted on a rubber disc can have its shape changed by locally varied stretching of the rubber."

³⁷ This paragraph explains once more the distinction between what space is as a form of intuition and what the axioms assert about space. Once again Helmholtz' account seems to us to demand the interpretation given in note 33. When we remove from spatial intuition everything that can be expressed conceptually, i.e. in the last analysis by geometrical axioms, there precisely remains only that qualitative element of spatiality (extendedness) which we witness as an ultimate datum not to be analysed further. Helmholtz did not state this himself, and it looks as if he did think of the concept of a form of intuition as endowed with a richer content. But since he nowhere explicitly stated what this content is, it is the task of interpretation to determine the latter such as appears compatible with the psychological and geometrical facts †.

---

† [This perhaps goes too far in attempting to 'rescue' Helmholtz. In fact rather than removing everything conceptual from intuition, Helmholtz locates his advance over Kant in having resolved the concept of intuition into elementary thought processes. See the subsequent course of the discussion.]

The most complete statement of what Helmholtz thought of as the content of pure spatial intuition is found – though admittedly only in the form of a comparison – in the following passage (*Wissenschaftliche Abhandlungen*, vol. II, pp. 641 f.), which is taken from the essay directed against Land (see below note 72):

"To recall a quite similar situation, it undoubtedly lies in the makeup of our visual apparatus that everything seen can only be seen as a spatial distribution of colours. That is our innate form of visual perceptions. But this form in no way prejudices how the colours we see are to be spatially ordered one beside another or to follow one another temporally. In the same sense, in my opinion, our representing all extenal objects in spatial relationships could be the only possible and *a priori* given form in which we can represent objects at all, without this needing to impose any constraint whereby after or beside certain specific spatial perceptions some other specific one must occur, so that e.g. every rectilinear equilateral triangle will have angles of 60 degrees however great its sides may be.

"In Kant, of course, the proof that space is an *a priori* given form of intuition relies essentially upon the belief that the axioms are synthetic propositions given *a priori*. Yet even if one eliminates this proposition and the proof based upon it, the form of spatial intuition could nevertheless still be given *a priori* as the necessary form of intuition of the existence one beside another of different things. In this no essential feature of Kant's system would be lost. On the contrary, this system would gain in consistency and understandability, because there would then also be eliminated the proof, constructed essentially upon the persuasive power of the axioms of geometry, for the possibility of a metaphysics – of which science Kant himself of course did not know how to discover anything more than the axioms of geometry and natural science. As regards the latter, they are partly of disputed correctness and partly simple inferences from the principle of causality, in other words from the urge of our faculty of understanding to consider everything that occurs to be lawlike, i.e. comprehensible. But since Kant's critique is otherwise everywhere directed against the admissibility of metaphysical inferences, it seems to me that his system has been freed from an inconsistency, and that a clearer concept of the nature of intuition has been attained, when one gives up the *a priori* origin of the axioms and regards geometry as the first and most perfect of the natural sciences."

[38] This description does not wholly do justice to Kant's theory of intuition, since it pays no attention to 'pure' intuition, which according to Kant is displayable in empirical intuition as the latter's form and lawlikeness, and itself is not a 'psychic process'. Even for Kant it was further resolvable, inasmuch as it indeed splits up into the individual geometrical axioms. Thus when Helmholtz charges Kant with not having tried to resolve intuition further, he must mean something else. He obviously wants to say that Kant failed to ask the question *why* spatial intuition contains in itself precisely the axioms which in fact hold, and not other ones. One must also interpret in this sense the passage in Appendix II (p. 150) where Helmholtz says: "The question of whether intuition was more or less resoluble into conceptual constructions had at that time not yet been raised."

[39] Compare the lecture on the origin and significance of the axioms of geometry, p. 6.

[40] This sentence has no parallel in the lecture on the axioms, and it raises a significant issue. We show elsewhere (note I.38) that in the case of non-Euclidean geometry the exclusion of 'every other interpretation' can strictly speaking never come about by logical compulsion, but instead only by cognitive economy, since besides the geometrical interpretation the physical one always remains possible from a purely theoretical viewpoint.

A full discussion of this matter is found in Helmholtz' repudiation of Land's criticism, which is reprinted in this volume (Appendix III).

[41] Concerning the expression "metamathematical" Helmholtz says elsewhere (*Wissenschaftliche Abhandlungen*, vol. II, p. 640): "The name was of course bestowed in an ironical sense by opponents, and modelled upon metaphysics. But since the developers of non-Euclidean geometry have never maintained its objective truth, we may very well accept the name."

[42] Helmholtz is wholly justified in construing the concept of intuitability, for the present purpose, precisely as is done here, since it is distinct only in degree and not in principle from what he in the next paragraph calls the 'older concept of intuition'. In these philosophical considerations, one is naturally dealing purely and simply with determining *in principle* the concept of intuition. A thing must count as accessible to intuition, if one can formulate methods with whose help it could be made representable to us in a sensory manner. A person born without eyes cannot in any way learn what a sighted person senses in perceiving yellow: for him there is thus no such method. Or if it were reported that some or other living things possessed a sense organ unknown to us – say for magnetic disturbances – we could in no way whatsoever procure ourselves an experience of the corresponding sensations. These are examples of cases lying beyond all intuitability. The next part of Helmholtz' account explains that the intuitability of non-Euclidean spaces does not belong to cases of that kind, but instead only requires depicting a succession of perceptions which are put together from purely everyday sensations.

[43] The situation may be illustrated with an example often used by Helmholtz elsewhere. One can form the following inference:

> Major premiss: Light which I see with my right eye in the vicinity of my nose, originates from a light source lying to my left.

> Minor premiss: When a certain pressure occurs on my right eyeball, I see a patch of light in the direction of my nose.

> Conclusion: The source of the light sensation lies to my left (i.e. the pressure lies to my left).

The inference (as is well known, one calls an inference of this form a 'syllogism') is false, since it is well known that the eye must be pressed on the right hand side in order that the arising semblance of light should appear to be localised to the left. The mistake arises through one's falsely considering the major premiss 'formed from a series of experiences' to be universally valid, and applying it to a case where it does not hold.

In his *Physiologische Optik* [*op. cit.*], 2nd ed., p. 582, Helmholtz says: "In millionfold repeated experiences, throughout our whole life, we have found that when we felt an excitation in the nervous apparatuses whose peripheral ends lie on the right hand sides of our two retinas, there lay a luminous object before us to our left. We had to raise our hand to the left in order to mask the light or to grasp the luminous object, or we had to move to the left in order to approach it. Thus although a genuine conscious inference is not present in these cases, the essential and primary task of one is accomplished and its result achieved, if admittedly only through the unconscious processes of the association of representations. This association goes on in the obscure background of our memory, and its results therefore also force themselves upon our consciousness as if obtained by way of a compelling, seemingly external power, over which our will has no authority. Of course these inductive inferences, which lead to the formation of our sense perceptions, lack the work of purification and examination carried out by conscious thought. As regards their proper nature, nevertheless, I believe I may still term them *inferences* – unconsciously performed inductive inferences."

[44] On the celebrated theory of 'unconscious inferences', which is surveyed in a few words in the paragraph just finished, we shall just briefly make the following comments. Modern psychology energetically rejects the concept of unconscious inference, because it rightly considers thought – the logical process – to be exclusively a function of *consciousness*. It may be asked whether Helmholtz merely uses an unsuitable terminology, or whether the improper terminology is also the expression of thoughts which do not stand up to rigorous epistemological criticism. We believe that Helmholtz' account, within broad limits, allows the first and favourable interpretation and therefore in fairness calls for it.

In the present paragraph Helmholtz abandons the term 'unconscious inferences', admittedly only in order to avoid confusions with Schopenhauer's 'wholly unclear' thoughts. Otherwise he had no substantial reservation about retaining the expression, as emerges from the fact that the passage given in our previous note was taken over unaltered by him from the first edition of his *Physiologische Optik* [*op. cit.*] in the second edition (1894), and that he expressly added precisely there (p. 602) that he "even now finds the name, up to a certain limit, to be still admissible and significant."

However, he plainly states that in actuality the process consists of "processes of association", and also pronounces similarly in other passages, e.g. p. 601 of the 2nd ed. of his *Physiologische Optik*, where he says: "So we see that although this process in its essential parts is brought about – as far as we can discern – only by an involuntary and unconscious action of our memory, it is capable of producing in us representational combinations whose outcome concurs in all essential features with that of conscious thought." Helmholtz justifies carrying over the logical term to these psychic processes by noting that the latter lead to like *results*, thus render the same services as proper inferences would. Associations and instincts undoubtedly even guide us, in general, more surely in our behaviour towards our environment than does our faculty of understanding. Of course, an inference from like achievement to like nature would not yet on that account be admissible. According to all this, one may well say concerning the questions broached here – where it is a matter of the order of our perceptions – that one need contest only the formulation of Helmholtz' theory of 'unconscious inferences' , and not what properly lies at the heart of it.

We shall not investigate here whether this also holds when Helmholtz wants to make unconscious inference responsible not merely for the order of perceptions, but also for the assumption of the existence of bodies as causes of sensations. He does this e.g. in the lecture 'Über das Sehen des Menschen' ['On human vision'] which was held to be the best of the Kant memorial in 1855 in Königsberg. Here he says (*Vorträge und Reden*, 5th ed., vol. I, p. 112): "But if consciousness does not perceive bodies immediately at the places of these bodies themselves, it can only come to an acquaintance with them by an inference." (Similarly Schopenhauer in *Die Welt als Wille und Vorstellung* ['The world as will and representation'], §4: "The first, simple, constantly available expression of the understanding is the intuition of the actual world; this is altogether a knowledge of the cause from the effect...,") But since Helmholtz later expressed himself more cautiously on this point (see towards the end of the address on the facts in perception), and since we are concerned not with the historical development of Helmholtz' views, but with their material and continuing significance, the earlier formulation can be left out of consideration.

More detail on Helmholtz' theory of unconscious inferences is to be found in B. Erdmann's repeatedly cited academy essay and in F. Conrat, *H. v. Helmholtz' psychologische Anschauungen* ['H. v. Helmholtz' psychological conceptions'], Halle, 1904, ch. 8.

[45] Lotze regards as local signs of this kind the eye movements (sensations of movement) which are necessary in order to bring the relevant place on the retina to the position of clearest vision. According to W. Wundt (*Physiologische Psychologie* ['Physiological psychology'], 5th ed., vol. II, pp. 668 ff.), the role of local signs could be played e.g. by differences of colour as well. It is indeed well known that the peripheral parts of the retina e.g. have a quite different colour sensitivity from that of the central ones.

[46] This objection was raised e.g. by du Bois-Reymond after reading the lectures which Helmholtz had given under the title 'Die neueren Fortschritte in der Theorie des Sehens' ['The recent advances in the theory of vision'], and which were reprinted first in the *Preussische Jahrbücher* and then in *Vorträge und Reden*, 5th ed., vol. I, pp. 265ff. In 1868 (see Koenigsberger, *H. v. Helmholtz*, vol. II, p. 84) he wrote to Helmholtz: "It seems to me still to speak against the strictly empiricist conception, that it precisely should be implementable consistently throughout, which – as you concede yourself – is not the case. For if it is innate in the baby calf to go towards the udder on account of the smell, what else might not then be innate in it? To me there seems to remain still so much ineliminable nativism that a bit more or less makes no difference...."

[47] The preceding paragraphs contain a description of the empiricist theory of visual perceptions, as advocated and thus named by Helmholtz. In this description he contrasts its fundamental thoughts clearly and brilliantly with the nativist theory, so that on occasion it might be said (by Fr. Hillebrand in his essay on Ewald Hering, Berlin, 1918, p. 102) of this famous passage that it could "count as the most perfect portrayal of the empiricist theory."

This is not the place to go into the controversy between the nativist and empiricist views, which even at the present has not exhausted itself. For this is a matter of questions belonging purely to a particular science, and which make no difference as regards the epistemological problem situation. It is true that Helmholtz says (*Physiologische Optik* [*op. cit.*], 1st ed., p. 796; 2nd ed., p. 945). "In their choice from the various theoretical views, it seems to me... that the various enquirers have been influenced more by a tendency to certain metaphysical ways of regarding things than by the pressure of facts", and one actually can at times ascertain an influence of the philosophical upon the physiological viewpoint. But it is neither necessary nor justified. The empiricist conception of spatial perception is quite independent of epistemological empiricism, as one may already infer from the fact that e.g. E. Mach combines a far-reaching nativism with a rigorously empiricist – indeed sensualist – epistemology. In fact one cannot see why the nativist assumption, that sensations possess in advance certain spatial properties, should not be just as much compatible with epistemological empiricism as is the fact – which even the latter of course concedes – that sensations are imbued with a quality and modality which is basic and cannot itself be derived.

It must also be stressed emphatically that it is an error to regard nativism as a form of Kantian apriorism, as still sometimes occurs. The purport of Kant's theory is to explain the apodictic validity of the axioms of geometry (compare note I.3). But nativism is a theory of *sense perception*, its aim cannot then be to give a foundation for a *rigorous* mathematical lawlikeness of space, since all perception as such supplies only approximate data. Thus for example v. Kries quite correctly says (in the 3rd ed. of Helmholtz' *Physiologische Optik*, vol. III, p. 524) that Hering's nativism "no longer has anything at all in common with Kant's conceptions".

Besides, even within psycho-physiology the opposition of nativist and empiricist conceptions is probably not as fundamental and irreconcilable as is sometimes assumed. A nativist theory which on the issue of localisation does not acknowledge a considerable role to experience as well, seems to be as incapable of implementation as an empiricist one wanting to undertake a construction of spatial representation from purely non-spatial elements of sensation. We also believe that Helmholtz' theory is not to be conceived as an attempt of this latter kind. On the contrary, he indeed considered the spatial to be a form of intuition of a basic kind (in his psychological sense), and we were obliged to interpret this (notes 33 and 37) to mean that every elementary sensation has a component of extendedness (with visual sensations perhaps 'surfacelikeness') which did not arise only at a subsequent stage through experience. Helmholtz' sentence "Everything our eye sees, it sees as an aggregate of coloured surfaces in the visual field – that is its form of intuition" (p. 129) seems in this respect to be wholly conclusive (compare also the quotation given in note 37). As to whether it might prove necessary at some point or another to come still nearer to nativist conceptions, this – as already said – makes no difference at all to the purely epistemological problem of perception.

[48] The task of founding a belief in the existence of the external world upon experiences of the will has been undertaken, in a somewhat different manner, especially by W. Dilthey. See 'Beiträge zur Lösung der Frage vom Ursprung unseres Glaubens an die Realität der Aussenwelt und seinem Recht' ['Contributions to solving the question as to the origin of our belief in the reality of the external world and as to its justness'], *Sitzungsberichte der Berliner Akademie*, 1890.

[49] The most extreme subjective idealism, which denies the existence of an external world different from the subject, is called (as is well known) solipsism. The theoretical possibility of this viewpoint has been conceded by the majority of philosophers, although it has naturally found no serious advocates. This passage indicates that the realist manner of expression in which Helmholtz has clothed his theory of perception does not emanate from an uncritical manner of thought, but is merely chosen as the formulation which most strongly suggests itself, while the realist or idealist interpretation of this formulation can be left to the metaphysician.

[50] Compare e.g. Fichte, *Bestimmung des Menschen* ['The determination of man'], book III, p. I.

[51] In this case the existence of the external world would have to be termed a *fiction* – a thought which H. Vaihinger has developed further in several passages in his *Die Philosophie des Als Ob* ['The philosophy of as if']. The present passage in Helmholtz seems to have escaped him, since he does not make reference to it in his book. For Helmholtz himself the realist view of the world is naturally not a fiction, but precisely a *hypothesis*.

[52] In this (probably not very apt) extended sense, the term 'thinking' covers all processes of the mind which in some way or another lead to appropriate behaviour towards the external world, and above all its associative and reproductive activities.

[53] The thought expressed here by Helmholtz is a fundamental insight of all epistemology. It lies at the base already of Plato's theory of Ideas – since Plato's Ideas are the unvarying paradigms of the eternally alternating individual things – and likewise at that of the modern theory of science. Stanley Jevons begins his *The Principles of Science* with the sentence: "Science arises from the discovery of Identity amidst Diversity."

[54] The sentence should not be understood literally, since a *law* naturally cannot at all be the object of a perception. Establishing something lawlike is rather always only the termination of a process of observing, ordering and interpreting. Helmholtz only wants to say that detecting laws is something more immediate than ascertaining substances, and it is this thought which is developed in more detail in the next three sentences. If one pursues this thought further, one easily arrives at an insight which is revealing itself more and more plainly in modern natural science, namely that the concept of substance can indeed be wholly reduced to the concept of a law, thus that on the highest level of cognition of nature the former is dispensable as a basic concept.

[55] The surprising manner in which Helmholtz here wishes to define the concept of a cause, by identifying cause and law[†], would surely be truly inappropriate. For one obviously lacks the right to retain the word 'cause', if one does not use it to characterise a concept concurring at least in its chief features with what one otherwise usually understands by it. If the causes are to be what eternally remains the same, while the effects – the happenings in the world – unremittingly alternate, then one will wholly destroy the reciprocal correspondence between the two, which otherwise belongs to their concepts; an effect could no longer itself be conceived of as the cause of further effects. One cannot see why we should introduce for the concept of a *law* the word cause in addition, which otherwise has a different sense, nor why it should be inappropriate to use the words cause and effect solely for processes in nature, to which their meaning is otherwise restricted in careful scientific usage. It is correct that the term cause is often used in 'a very wishy-washy manner'; but that can be avoided without there having to arise a confusion with the concepts of the antecedent or of the occasion of something. Compare Schlick, *Naturphilosophische Betrachtungen über das Kausalprinzip* ["Reflections on the principle of causality from the point of view of natural philosophy"], *Die Naturwissenschaften*, VIII, pp. 461 ff.

[56] This pronouncement might easily be misunderstood without more detailed interpretation. It is probably based in the first place on the thought that if – by a kind of anthropomorphism – we take the concept of a cause (which Helmholtz has indeed just identified with that of a law) to be that of something "compelling the course of natural processes, to be a power equivalent to our will", then it does indeed turn into the concept of force.

In order to characterise Helmholtz' viewpoint further, we may adduce some statements from §5 and §6 of his *Einleitung zu den Vorlesungen über theoretische Physik* ['Introduction to the lectures on theoretical physics']. There he explains how say the law of attraction initially only reads "Two heavy bodies at a finite distance from each other in space undergo an acceleration, and indeed each of them in the direction of the other", but how we then "by forming abstractions and substituting nouns for the verbs... express it in the form that between any two heavy bodies... there exists continuously a force of attraction of a certain magnitude. We have thereby introduced, in place of the simple

---

[†] [To be exact, Helmholtz does not intend 'cause' and 'law' to be synonyms, but reserves the former for a very securely established law.]

description of the phenomenon of motion, an abstraction – the force of attraction. We thereby indeed signify nothing – at least nothing still having a factual sense – beyond what is also contained in the description of the mere phenomenon. In laying down the law in this form, which uses the concept of force, we merely add an assurance that this phenomenon of mutual approach of the two bodies occurs – as soon as the conditions for it are given – at any moment of time."

[57] It is certainly surprising that Helmholtz quotes Fichte relatively frequently and always assentingly, although the rationalist system of this thinker indeed hardly displays any points of contact with his views, and although the form in which Fichte expounded his theoretical ideas certainly could not be to his taste. One can find an explanation only in the fact that Helmholtz, in appraising Fichte, let himself be lead by feelings of piety. His sympathy for the philosopher undoubtedly goes back to impressions which he received, according to his own account (*Vorträge und Reden*, 5th ed., vol. I, p. 17), already as a boy in the house of his father, who was an enthusiastic admirer of Fichte.

[58] These famous words of Kirchhoff, which have as it were become a programme for epistemologically oriented physics, are elucidated by him himself (in the foreword to his lectures on mechanics) as follows: "I wish by that to say that it should be a matter only of stating *what* are the appearances that occur, and not of ascertaining their *causes*." He saw himself forced into this position because the previously customary definition of mechanics was not satisfactory: "One usually defines mechanics as the science of forces, and forces as *causes* which bring about motions or *strive* to bring them about," and this definition "is infected with the unclarity which cannot be eliminated from the concepts of a cause and of striving." Kirchhoff's motive was thus to get away from anthropomorphism, which we spoke of above (note 56).

[59] Let us briefly look at what Helmholtz teaches, in this and the preceding paragraph, about the content and validity of the principle of causality. *Contentually* it expresses, in his (undoubtedly correct) opinion, a trust in the complete comprehensibility of the world; and since this is identical with a trust in all-pervading lawlikeness, "the principle of causality is in fact nothing but the presupposition of the lawlikeness of all of the appearances of nature" (as Helmholtz put it in an addition written in 1881 to his paper on the conservation of force). As regards its *validity* it is a *regulative* principle, which thus serves as a guide line for enquiry, but for which we have no further guarantee than its success. Thus its validity is a factual one in which we must trust, but which we cannot prove.

It is quite different with Kant. For him the law of causality is a *constitutive*[†] principle,

---

[†] [This is an astonishing statement, since Kant expressly states that the principle of causality is a regulative principle, see *Kritik der reinen Vernunft* [*op. cit.*], 2nd ed., pp. 221–3. (This is in fact the source of the distinction between constitutive and regulative principles.) However, Kant means something different from what Helmholtz does by a regulative principle. For Kant this means, in the case of the principle of causality, that we know *a priori* that any given event necessarily has a cause, but do not necessarily know *a priori* what that cause is – this in general has to be ascertained empirically. While an example of a constitutive principle for Kant is that every sensation we experience must have some degree of intensity and that the possible degrees form a continuum; this is a constitutive principle, because it enables one to determine *a priori* what a stronger intensity of a given sensation will be by combining a sufficient number of weaker intensities of that sensation.]

it has a share in the construction of our experience and first makes it possible; it must there-fore be necessarily valid for all experiences. According to Kant we do not merely trust in its validity, but are assured of it.

⁶⁰ Once again Helmholtz uses the words *a priori* and transcendental in a quite different sense from what Kant does. According to the philosopher, a proposition is called *a priori* if it is *valid*, and can be seen to be valid, independent of experience. And precisely this, according to what Helmholtz declares both before and after this point, is not so with the principle of causality. In using the words *a priori* Helmholtz wants merely to state that the principle of causality cannot be gathered from experience by induction, but instead must always already be presupposed in the interpretation of experiences. But such a pre-supposition, whose validity is not established in advance, has the character of a *hypothesis*.
    Helmholtz also became clear about this, for in a note which he left behind (Koenigs-berger, *H. v. Helmholtz*, vol. I, pp. 247f.) he says: "The law of causality (the presupposed lawlikeness of nature) is only a hypothesis, and cannot be proved to be anything other than this. No lawlikeness to date can prove a future lawlikeness. As contrasted with other hypotheses, which express particular laws of nature, the law of causality has an exceptional status only as follows: 1. It is the presupposition for the validity of all the others. 2. It gives us the only possibility whatsoever of knowing something which is un-observed. 3. It is the necessary basis for conduct having a purpose. 4. We are driven to it by the natural mechanics of our representational combinations. Thus we are driven to *wish* it to be correct by the strongest motives; it is the basis of all thought and conduct. Until we have it, we cannot even test it; thus we can only *believe* in it, *conduct* ourselves according to it...." Helmholtz also makes reference to conduct, to the *practical* proof of the principle of causality, in the immediately following statements in the text.
˙ Thus it is not Kant's viewpoint which he advocates on the issue of the law of causality, but instead he takes the path of David Hume. We may be allowed to add that to us too Hume's viewpoint seems to be the only one which can withstand all critical attacks.

⁶¹ Contrary to this assurance, it has emerged in our critical comments – and has also indeed already been stressed quite often – that only a few traces of philosophical agreement can be ascertained between Kant and Helmholtz. The doctrine of the subjectivity of spatial intuition and the qualities of the senses, which admittedly was the most important thing for Helmholtz, is properly the only point on which he could justly and unrestrictedly make an appeal to Kant. The explanation why Helmholtz himself believed in a greater agreement than in fact existed is to be found partly in his not always grasping Kant's doctrine cor-rectly, but instead interpreting it too much in a psychological sense, and partly in that his high esteem for that thinker made the things in common appear to him more important, and the deviations less essential.

⁶² In the text, Helmholtz has advocated the opinion that a spatial interpretation is given only to those data of consciousness which alter when body movements are executed. Here he raises against himself the objection that many sensations, originating in the interior of the body, are not noticeably influenced by movements, yet that we nonetheless do not take them to be non-spatial psychic states (like memories or wishes), but rather localise them more or less distinctly at specific places in our bodies. In what follows, the arguments which he brings against this objection are thoroughly suited to removing its force, or at least blunting its edge, and Helmholtz here also has modern psychology essentially on his side. Admittedly, the nativist view that internal sensations are in a certain manner localised

in advance – thus without movements being involved – is perhaps one that also cannot be regarded as wholly refuted and impossible.

⁶³ It is notable that Helmholtz classes states like depression and melancholy amongst sensations, whereas otherwise they are of course distinguished from them as being states of *feeling*. He thereby seems to anticipate the chief thoughts of the ingenious theory of feelings given by W. James (1884) and K. Lange (1887), according to which feelings are supposed to consist of nothing other than sensations of internal organs.† The James-Lange theory is admittedly rejected by most modern psychologists.

⁶⁴ The uncertainty of localisation for the sensations of internal organs is also reduced by E. Meumann (*Archiv für die gesamte Psychologie* **9** (1907), 57) in the first place to the fact that we cannot *see* the organs, thus cannot follow their motions.

⁶⁵ In previous notes (33, 37) it already plainly emerged in what sense we must understand Helmholtz' words in order for the claim made in this heading to be able to count as correct We came to the following result. If 'transcendental' – in accordance with Helmholtz's linguistic usage – means the same as *a priori*, and if the latter is understood in Kant's meaning, then the *a priori* nature of the axioms is no different from the *a priori* nature of space. In Kant the assertion of the *a priori* nature of the latter has indeed only the purpose of explaining the apodictic validity of the axioms. But Helmholtz' *a priori* is precisely not Kant's, and instead has only the meaning that the spatiality of perceptions is something purely subjective in the same sense as are the qualities of sensations (compare note 16). Under this presupposition, of course, the 'transcendental' nature of spatial intuition need yield no ground for the validity of any synthetic *a priori* axioms about space.

⁶⁶ Erdmann in fact there (p. 97) declares spatial intuition to be *a priori* inasmuch as it could not arise in us "unless there were precisely a particular disposition of our psychic activities which grasps in a spatial manner certain groups of external stimuli." But it must "in other respects equally justly" be called empirical, meaning obviously: inasmuch as its content can be expressed in geometrical axioms. These rest upon experience, since "there must for the extension of our space in three dimensions and its unboundedness, as also for its continuity, and lastly for the particular value of its measure of curvature, be particular occasioning factors in the excitations evoked by the psychic development of spatial representation." In these pronouncements the *a priori* nature (or "transcendentiality") of spatial intuition is obviously taken in the same meaning as in Helmholtz himself.

⁶⁷ Since Krause adopted Kant's standpoint (which does not mean to say that he interpreted the philosopher's doctrines correctly in every detail), he was permitted to treat the two questions – of whether space is an *a priori* form of intuition and of whether the axioms of geometry are synthetic judgements *a priori* – as having the same meaning. His fault indeed was that in all of his presuppositions he already placed himself in Kantian territory, and undertook to criticise Helmholtz from that starting point. Since an insight into the new theory of space is only possible precisely when one frees oneself from these presup-

---

† [In fact what Helmholtz states in this paragraph is that such feelings may be *falsely* localised in the body because their causes need not always be psychic, i.e. he in effect rejects the James-Lange theory.]

positions, he was completely devoid of understanding in his stance towards the true content of the Riemann-Helmholtz theory. It is not worth going into Krause's arguments in more detail at this point than Helmholtz himself does in the text.

[68] Compare note 47.

[69] Compare note 38.

[70] Regarding the fact that this belief always remains *possible*, but is not justified, see above notes I.38 and I.49.

[71] We had already drawn attention to this earlier (note I.25). Helmholtz censures a misunderstanding of this name which has not yet been rooted out even today, since in popular articles directed against the theory of relativity one still finds, again and again, the assertion that it is a mistake to speak of 'curved' spaces.

[72] The considerations which Helmholtz here adds as Appendix III were first published (in an English translation) in the English periodical *Mind*, April 1878, no. X, as an answer to criticism expressed in *Mind*, no. V by Professor Land on Helmholtz' paper on the origin and significance of the axioms of geometry. In the paper in *Mind* (reprinted in his *Wissenschaftliche Abhandlungen*, vol. II, pp. 640–660 under the title 'Über den Ursprung und Sinn der geometrischen Sätze. Antwort gegen Herrn Prof. Land' ["On the origin and sense of the axioms of geometry. An answer to Prof. Land"]), there are some initial paragraphs which Helmholtz here left out, as their material content is already contained in another form in 'The Facts in Perception' . For the same reason they could remain absent in this edition. A short extract from them has already been reproduced above (note 37). The content of this appendix is highly important for Helmholtz' train of thought, as it fills a conspicuous gap which had been left in the proof given in the text. Compare note I.49.

[73] Compare the text, p. 137.

[74] These two kinds of alikeness are also contrasted by Helmholtz elsewhere (*Wissenschaftliche Abhandlungen*, vol. II, p. 641) by terming the first "*subjective* alikeness in the hypothetical transcendental intuition", and the second "*objective* alikeness in value of the real substrates of such spatial magnitudes, which verifies itself in the course of physical relationships and processes."

[75] In this definition, Helmholtz reduces the "physical alikeness in value" of spatial magnitudes to alikeness of physical conditions and alikeness of periods of time and processes. One must ask whether a circular definition is not contained in this, for how will one ascertain the alikeness of the mentioned circumstances – e.g. of the temporal magnitudes – other than with the help of spatial alikeness, since all physical methods of measurement (instrument readings) surely reduce ultimately to the latter? The formulation therefore cannot satisfy us, and only the next two sentences in the text will let the correct and usable sense of Helmholtz' thought emerge.

[76] Here Helmholtz founds the concept of physical alikeness in value upon specific facts of experience, and in this way augments his above definition. On looking more closely,

one finds that the experiences involved here always ultimately come down to observing coincidences of material points. Compare note I.39.

[77] Compare note I.54. We earlier distinguished physical geometry from purely conceptual geometry. The third kind of geometry which might at first seem possible – the *a priori* geometry of pure intuition – is precisely rejected by the considerations which Helmholtz will now give.

[78] The two triangles *ABC* and *Abc* would not be similar to each other in a non-Euclidean geometry. It was already remarked above (note I.17) that there are similar figures only in Euclidean geometry.

[79] Compare the lecture by F. Klein cited earlier (note I.49).

[80] In fact, 'pure intuition' is according to Kant the embodiment of a fully rigorous lawlikeness; it is after all supposed to yield the ground of the inviolate validity of the axioms. Visual estimation, on the other hand, would in Kant's terminology presumably have to be counted somehow as 'empirical' intuition. How exactly pure is related to empirical intuition is something about which we are able to find no wholly satisfying statements in Kant. But at any rate, in his opinion pure intuition finds its expression in empirical intuition by specifying the latter's lawlikeness – its 'form' – in some manner or another. The possibility of confusing the two of them seems not to be excluded with sufficient certainty to prevent statements of empirical intuition – thus of experience – from creeping into the arguments of Kantians as statements of "pure" intuition.

[81] In this case, the proposition that the two kinds of alikeness are identical would assert that with the second triangle too the sides *must* turn out to be alike on measuring with a pair of dividers. But only experience (according to Kant as well, as Helmholtz correctly emphasises) can tell whether, for the stated construction, a specific divider tip will coincide with a specific point in our diagram. Should the result of the measurement not agree with our Euclidean expectations, the reason could lie either in a deformation of the likened objects or in the unserviceability of Euclidean geometry. The Kantian rejects the latter possibility in advance, but – as Helmholtz demonstrates – without sufficient reason.

[82] It may be appropriate to sum up briefly the argument of the last four paragraphs. Helmholtz reasons as follows: suppose that what the Kantians think were actually so, i.e. that from the two interpretations possible in principle (see the previous note, also note I.49) we were always constrained by some pure intuition to choose the physical one which leaves Euclidean geometry untouched. Then if bodies behaved in a manner which Helmholtz describes as a behaviour in a non-Euclidean world, we would admittedly have to conceive of this behaviour as a peculiar motion of objects, accompanied by strange deformations, in Euclidean space; but we could ascertain, by conceptual-analytical methods, that all natural processes proceeded *as if* the world had a non-Euclidean constitution, thus that the world could be most simply and perspicuously described with the help of non-Euclidean metrics. And we should have to feel the constraint of intuition – a constraint which would like to forbid us precisely this manner of portrayal – as an irksome obstacle blocking our way to the logically most convenient conception, and concealing the latter's simple kernel with a complicated husk. Led by the principle of economy, we could indeed arrive at the assertion that the 'actual' constitution of the world is non-Euclidean; while

the sensory conception which – under the presuppositions made – even the enquirer who knows would naturally be unable to eliminate, might not unjustly be termed by him a kind of 'false semblance' .

[83] As is well known, modern physics has gone over in the general theory of relativity to the assumption of non-Euclidean metrics, it ascribes to space a measure of curvature which varies from place to place. The situation described in the preceding note has thereby actually occurred: our Euclidean manner of representation is in combat with a conceptual analysis which discerns that Euclidean intuition is not wholly adequate to actuality. But it holds us in its fetters, thus e.g. we call light rays which go past the sun *curved*, and portray them thus in diagrams, although these are of course the straightest lines in our space.

The factual situation only fails to correspond to the assumptions of our preceding note, to the extent that the deviations from Euclidean conditions are so small that they do not reveal themselves to everyday sense perception. If they did this, however, we are convinced like Helmholtz that our intuitive faculty of representation would then be completely adapted to the non-Euclidean metrics, because it would precisely have developed under the influence of non-Euclidean experiences. Thus there would then be no conflict at all between the intuitive representation and the conceptual portrayal of nature. As regards the situation which in fact exists we may thus infer: the difficulty which representing Einstein's world gives to most people is not to be reduced to an inescapable *a priori* form of intuition of human consciousness, but to the fact that all of our everyday experiences lead to a physical geometry which is Euclidean to within the accuracy of perception for our senses.

[84] Thus for the idealist the topogenous and hylogenous factors would have to be sought in some or other properties of our perceptions or other data of consciousness, since he of course does not acknowledge any reality apart from them. Nothing is presupposed at all about these factors, thus not even say that the two kinds are wholly independent of each other. Only experience can tell whether this is the case. It is well known that according to the general theory of relativity, 'material' and spatial specifications are not separable from each other.

[85] The view which Helmholtz here criticises as the opinion of many Kantians is one with which Kant himself cannot be taxed. See the passages from Kant which are cited in note 29 above.

[86] The expression 'real world' must be taken here in the sense in which it has a sense even for the subjective idealist. In his world too the topogenous and hylogenous factors of course exist, for him too there can be a 'form' of representation which would be recognisable as such a form in that its data are not accessible to testing by experience because every experience would already presuppose it.

[87] Compare here notes 23, 32 and 36 above.

BIBLIOGRAPHY

PART A. WORKS BY HELMHOLTZ

1. *Über die Erhaltung der Kraft: eine physikalische Abhandlung* (Berlin, 1847; Leipzig, 1862) 72 pp. Reprint in Ostwalds Klassiker No. 1 (Leipzig, 1899), with Helmholtz' *Zusätze* of 1881, pp. 53-60. Eng. tr. in *Scientific Memoirs*, ed. Tyndall and Francis (London, 1853) pp. 114–162.
2. *Handbuch der physiologischen Optik*, 3 vol. (Heidelberg, 1856, 1860 and 1867) 875 pp.; 2nd ed., in 8 parts (Hamburg, 1885–1894) 640 pp.; 3rd ed., 3 volumes, (Hamburg, 1909/1911/1910). Eng. tr. by J.P.C. Southall, *Treatise on Physiological Optics*, 3 vol., from the 3rd Ger. ed. for the Optical Society of America, (Rochester, N.Y., 1924–25; Dover reprint, 3 vol. in 2, N.Y., 1962) 482, 479 and 746 pp.
3. *Die Lehre von den Tonempfindungen als physiologische Grundlage für die Theorie der Musik* (Braunschweig, 1863) 605 pp. Second ed. 1865; 3rd, 1870; 4th, 1877; 6th, 1913. Eng. tr. of 3rd ed. by Alexander J. Ellis, *On the sensations of tone as a physiological basis for the theory of music* (London, 1875; 2nd ed. 1885; Dover reprint *Sensations of Tone* with intro. by Henry Margenau, 576 pp., N.Y., 1954).
4. *Populäre wissenschaftliche Vorträge*, 2 vol. (Braunschweig, 1865 and 1871; 2nd ed., 1876). The 5th ed. (1903) included:
   Vol. 1: 'Über Goethe's naturwissenschaftliche Arbeiten' (1853); 'Über die Wechselwirkung der Naturkräfte und die darauf bezüglichen neuesten Ermittelungen der Physik' (1854); 'Über die physiologischen Ursachen der musikalischen Harmonie' (1857); 'Über das Verhältniss der Naturwissenschaften zur Gesammtheit der Wissenschaften. Akademische Festrede' (1862); 'Über die Erhaltung der Kraft' (1862/63), 'Eis und Gletscher' (1865); 'Die neueren Fortschritte in der Theorie des Sehens' (1868); 'Über das Ziel und die Fortschritte der Naturwissenschaft. Eröffnungsrede für die Naturforscherversammlung zu Innsbruck' (1869).
   Vol. 2: 'Über den Ursprung und die Bedeutung der geometrischen Axiome' (1870); 'Zum Gedächtniss an Gustav Magnus' (1871); 'Über die Entstehung des Planetensystems' (1871); 'Optisches über Malerei' (1871–1873); 'Wirbelstürme und Gewitter' (1875); 'Das Denken in der Medicin' (1877); 'Über die akademische Freiheit der deutschen Universitäten' (1877); 'Die Tatsachen in der Wahrnehmung' (1878); 'Die neuere Entwickelung von Faraday's Ideen über Elektricität' (1881); 'Über die elektrischen Masseinheiten nach den Berathungen des elektrischen Congresses, versammelt zu Paris 1881'; 'Antwortrede, gehalten beim Empfang der Graefe-Medaille. Heidelberg, 9 August 1886'; 'Josef Fraunhofer. Ansprache, gehalten bei der Gedenkfeier zur hundertjährigen Wiederkehr seines Geburtstages. Berlin, 6 March 1887'; 'Goethe's Vorahnungen kommender naturwissenschaftlicher Ideen. Rede, gehalten in der Generalversammlung der Goethe-Gesellschaft zu Weimar 11 June 1892'; 'Heinrich Hertz. Vorwort zu dessen Principien der Mechanik. Berlin July 1894'.

   Partial Eng. tr. by E. Atkinson *et al.*, *Popular Lectures on Scientific Subjects* (London,

1881; 2nd ed., 1893), also includes the 1876 paper 'The origin and meaning of geometrical axioms'. The books are bound with the title *Popular Scientific Lectures*. They comprise:

Vol. one: 'On the Relation of Natural Science to Science in General' (tr. by H.W. Eve); 'On Goethe's Scientific Researches' (tr. by H.W. Eve); 'On the Physiological Causes of Harmony in Music' (tr. by A.J. Ellis); 'Ice and Glaciers' (tr. by E. Atkinson); 'On the Interaction of the Natural Forces' (tr. by Tyndall); 'The Recent Progress of the Theory of Vision' (tr. by Pye-Smith), 'On the Conservation of Force' (tr. by E. Atkinson), 'On the Aim and Progress of Physical Science' (tr. by W. Flight).

Vol. two: 'Gustav Magnus. In Memoriam', 'On the Origin and Significance of Geometrical Axioms', 'On the Relation of Optics to Painting', 'On the Origin of the Planetary System', 'On Thought in Medicine', 'On Academic Freedom in German Universities'.

The third German edition was entitled *Vorträge und Reden* (Braunschweig, 1884).

See also *Popular Scientific Lectures*, ed. by Morris Kline (Dover reprint, N.Y., 1962) with selections from the Atkinson translation: 'On Goethe's Scientific Researches'; 'On the Physiological Causes of Harmony in Music'; 'On the Interaction of Natural Forces'; 'The Recent Progress of the Theory of Vision'; 'On the Conservation of Force'; 'On the Origin and Significance of Geometrical Axioms'; 'On the Relation of Optics to Painting'.

5. *Wissenschaftliche Abhandlungen*, 3 vol. (Leipzig, 1882, 1883 and 1895) 938, 1021, 654 pp. Introductory essay: 'Hermann von Helmholtz' Wissenschaftliche Abhandlungen' by G. Wiedemann, 3, xi-xxxviii; and 'Titelverzeichniss sämmtlicher Veröffentlichungen von Hermann von Helmholtz' by A. König, 3, 605-636. For contemporary English translations of Helmholtz' papers, see the *Royal Society Catalogue of Scientific Papers* for 1869, 1872, 1877, 1894, and 1916.

6. *Schriften zur Erkenntnistheorie*, ed. with extensive notes by Paul Hertz and Moritz Schlick, (Berlin, 1921). Contains: 'Über den Ursprung und die Bedeutung der geometrischen Axiome' (1870); 'Über die Tatsachen, die der Geometrie zugrunde liegen' (1868); 'Zählen und Messen' (1887); 'Die Tatsachen in der Wahrnehmung' (1878 lecture in Berlin, with two supplements added later by Helmholtz). Tr. as *Epistemological Writings*, tr. M.F. Lowe, ed. R.S. Cohen and Y. Elkana (Dordrecht and Boston, 1977). This Volume.

7. *Helmholtz on Perception: Its Physiology and Development*, ed. with introductory essay by Richard M. Warren and Roslyn P. Warren (N.Y., 1968). Eng. tr. of: 'On the physiological causes of harmony in music' (Ellis tr. from Atkinson ed. above); 'The recent progress of the theory of vision' (Pye-Smith tr. from Atkinson above); 'On the relation of optics to painting' (Atkinson tr. from above); 'Concerning the perceptions in general' (Southall tr. from the *Treatise on Physiological Optics*, vol. 3, sect. 26); 'The facts of perception' (new tr. by Warren and Warren); 'The origin of the correct interpretation of our sensory impressions', from *Z. Psych. u. Physiol. Sinnesorgane* 7 (1894) pp. 81-96, later in conclusion of the *Physiological Optics*, 2nd ed. (1896) (new tr. by Warren and Warren).

8. *Selected Writings of Hermann von Helmholtz*, ed. Russell Kahl (Middletown, Conn., 1971), with a chronological bibliography of Helmholtz' works. The twenty selections are: 'The Conservation of Force: A Physical Memoir' (1847), [substantial rev. of Tyndall tr.]; 'The Scientific Researches of Goethe' (1853), [rev. of Eve tr. (in Atkinson

above)]; 'The Physiological Causes of Harmony in Music' (1857), [rev. of Ellis tr. (in Atkinson)]; 'The Application of the Law of the Conservation of Force to Organic Nature' (1861), [original in English]; 'The Relation of the Natural Sciences to Science in General' (1862), [substantial rev. of Eve tr. (in Atkinson)]; 'Recent Progress in the Theory of Vision' (1868), [rev. of Pye-Smith tr. (in Atkinson)]; 'The Aim and Progress of Physical Science' (1869), [substantial rev. of Flight tr. (in Atkinson)]; 'The Origin and Meaning of Geometric Axioms (I)' (1870), [rev. of Atkinson tr.]; 'The Origin of the Planetary System (1871), [substantial rev. of Atkinson tr.]; 'The Relation of Optics to Painting' (1871), [substantial rev. of Atkinson tr.]; 'The Endeavor to Popularize Science' (1874), [tr. Kahl]; 'Thought in Medicine' (1877), [substantial rev. of Atkinson tr.]; 'The Origin and Meaning of Geometric Axioms (II)' (1878), [original in English]; 'The Facts of Perception' (1878), [new tr. by Kahl]; 'The Modern Development of Faraday's Conception of Electricity' (1881), [original in English]; 'An Epistemological Analysis of Counting and Measurement' (1887), [new tr. by Kahl]; 'An Autobiographical Sketch' (1891), [substantial rev. of Atkinson]; 'Goethe's Anticipation of Subsequent Scientific Ideas' (1892), [new tr. by Kahl]; 'The Origin and Correct Interpretation of Our Sense Impressions' (1894), [new tr. by Kahl]; 'Introduction to the *Lectures on Theoretical Physics* (Introduction and Part I)' (1894), [new tr. by Kahl].

 9. *Counting and Measuring*, tr. by Charlotte T. Bryan, with an introductory essay, 'A survey of the problem of mathematical truth', by Harold T. Davis, (N.Y., 1930).
10. 'On Thought in Medicine', *Bull. Inst. Hist. Med.* 6 (1938) pp. 117–143, tr. with intro. by Arno B. Luckhardt.
11. 'The Origin and Meaning of Geometrical Axioms', I and II, *Mind* 1 (1876) pp. 301-321, and 3 (1878) pp. 212-225.
12. *Vorlesungen über Theoretische Physik*, ed. by Arthur König, Carl Runge, Otto Krigar-Menzel, and Franz Richarz, 6 vol. (Leipzig, 1897–1903): Band I, Abt. 1: *Einleitung zu den Vorlesungen über theoretische Physik* (1903) 50 pp.; Band 1, Abt. 2: *Die Dynamik discreter Massenpunkte* (1911) 380 pp.; Band 2: *Dynamik kontinuierlich verbreiteter Massen* (1902) 248 pp.; Band 3: *Mathematische Principien der Akustik* (1898) 406 pp.; Band 4: *Elektrodynamik und Theorie des Magnetismus* (1907) 406 pp.; Band 5: *Elektromagnetische Theorie des Lichtes* (1897) 370 pp.; Band 6: *Theorie der Wärme* (1903) 418 pp.
13. *Hermann von Helmholtz über sich selbst*, ed. Dorothea Goetz. Address on the occasion of the celebration of his 70th birthday, 1891 (Reprint, Leipzig, 1966; Eng. tr. in Kahl above).
14. 'On the interaction of natural forces', in *The Correlation and Conservation of Forces*, a remarkable mid-century collection of articles by Grove, Mayer, Faraday, Liebig and Carpenter in addition to Helmholtz, ed. by Edward L. Youmans (N.Y. and London, 1865) pp. 211-250.
15. 'Über die Entdeckungsgeschichte des Princips der kleinsten Action' and 'Zur Geschichte des Princips der kleinsten Action'. These two articles on the history of the principle of least action were published posthumously by Adolf Harnack in the bicentenary history of the Berlin Academy, *Geschichte der Königlich preussischen Akademie der Wissenschaften zu Berlin*, vol. 2 (Berlin, 1900) p. 287ff. They do not appear in the *Wissenschaftliche Abhandlungen*.

Note: Convenient reprint selections have been published in Germany, among them:

16. *Abhandlungen zur Thermodynamik*, Ostwalds Klassiker No. 124, ed. by Max Planck

(1921). Contains 'Über galvanische Ströme, verursacht durch Concentrations-Unterschiede: Folgerungen aus der mechanischen Wärmetheorie' (1877); 'Die Thermodynamik chemischer Vorgänge' (1882); 'Zur Thermodynamik chemischer Vorgänge' (1882 and 1883).

17. *Das Denken in der Naturwissenschaft* (Darmstadt, 1968). Four essays from *Vorträge und Reden*.

18. *Die Tatsachen in der Wahrnehmung. Zählen und Messen, erkenntnistheoretisch betrachtet.* (Darmstadt, 1959). The two essays of 1878 and 1887 as given in the Schlick-Hertz edition above.

19. *Über Geometrie* (Darmstadt, 1968). 'Über den Ursprung und die Bedeutung der geometrischen Axiome' (1870); 'Über die Tatsachen, die der Geometrie zugrunde liegen' (1868); 'Über den Ursprung und Sinn der geometrischen Sätze' (1878).

20. *Philosophische Vorträge und Aufsätze*, ed. Herbert Hörz and Siegfried Wollgast (Berlin, 1971). Fourteen papers:
'Erinnerungen'; 'Über Goethes naturwissenschaftliche Arbeiten'; 'Über das Sehen des Menschen'; 'Über das Verhältnis der Naturwissenschaften zur Gesamtheit der Wissenschaften'; 'Über die Erhaltung der Kraft'; 'Über das Ziel und die Fortschritte der Naturwissenschaft'; 'Über den Ursprung und die Bedeutung der geometrischen Axiome'; 'Das Denken in der Medizin'; 'Die Tatsachen in der Wahrnehmung', with Beilagen: (a) 'Der Raum kann transzendental sein, ohne dass es die Axiome sind'; (b) 'Die Anwendbarkeit der Axiome auf die physische Welt'; 'Zählen und Messen erkenntnistheoretisch betrachtet'; 'Goethes Vorahnungen kommender naturwissenschaftlicher Ideen'; 'Über das Streben nach Popularisierung der Wissenschaft'.

PART B. BIOGRAPHICAL MATERIALS

1. Koenigsberger, Leo, *Hermann von Helmholtz*, 3 vol. (Braunschweig, 1902–03) 375, 383, 145 pp. There is a slightly abridged English translation by Frances A. Welby, published in one volume (Oxford, 1906; Dover reprint, N.Y., 1965). Personal and intellectual biography, with expository precis of the scientific works, and copious quotation from letters and manucripts.

2. du Bois-Reymond, Emil, *Hermann von Helmholtz, Gedächtnisrede* (Leipzig, 1897) 80 pp. Also in his *Reden*, vol. 2 (Leipzig, 1912), ch. 37.

3. *Anna von Helmholtz: Ein Lebensbild in Briefen*, ed. Ellen von Siemens-Helmholtz, 2 vol. (Berlin, 1929).

PART C. WORKS ON HELMHOLTZ

Albring, Werner, 'Gedanken von Helmholtz über schöpferische Impulse und über das Zusammenwirken verschiedener Wissenschaftszweige', in Klare *et al.* (1972) pp. 7–23.

Alexander, Karl-Friedrich, 'Zum Begriff der Kraft bei Helmholtz', in Klare *et al.* (1972) pp. 65–68.

Bernfeld, Siegfried, 'Freud's scientific beginnings', *American Imago* 6 (1949) 163–196.

Bernfeld, Siegfried, 'Freud's earliest theories and the school of Helmholtz', *Psychoanal. Q.* 13 (1944) 341–362.

Boring, Edwin G., *A History of Experimental Psychology*, 2nd ed. (N.Y., 1950). Ch. 15 on Helmholtz.

Boring, Edwin G., *Sensation and Perception in the History of Experimental Psychology* (N.Y., 1942).

Brush, Stephen G., *The Kind of Motion We Call Heat: A History of the Kinetic Theory of Gases in the 19th Century*, 2 vol. (Amsterdam and N.Y., 1976).

Buchdahl, Gerd, *Metaphysics and the Philosophy of Science* (Oxford, 1969). A penetrating treatment of Kant, and the classical background.

Cappeletti, Vincenzo, 'Helmholtz heute', *Studia Leibniziana* 1 (1969) 253–262.

Cassirer, Ernst, *The Problem of Knowledge* (New Haven, 1950). Sub-titled 'Philosophy, science and history since Hegel', this is vol. 4 of Cassirer's *Das Erkenntnisproblem in der Philosophie und Wissenschaft der neueren Zeit*, and was first published in the English translation of W. H. Woglom and C. W. Hendel.

Cassirer, Ernst, *Determinism and Indeterminism in Modern Physics*, tr. O. T. Benfey (New Haven, 1956; 1st German ed., Göteborg, 1936). Especially concerned with Helmholtz on causality.

Cassirer, Ernst, *Substance and Function*, tr. W. C. and M. C. Swabey (Chicago, 1923; 1st German ed., Berlin, 1910). Especially concerned with Helmholtz on perception and the theory of signs.

Classen, J., *Vorlesungen über modernen Naturphilosophen* (Hamburg, 1908), 180 pp. Studies of DuBois-Reymond, F. A. Lange, Haeckel, Ostwald, Mach, Boltzmann, Poincaré, Kant and Helmholtz.

Conrat, Friedrich, 'Hermann von Helmholtz' psychologische Anschauungen', *Abhandlungen der Philosophie und ihrer Geschichte, Heft 18* (Halle, 1904), 278 pp. An expanded inaugural dissertation at Bonn.

Cranefield, P. F., 'The organic physics of 1847 and biophysics today', *J. Hist. Med.* xii (1957).

Cranefield, P. F., 'The philosophical and cultural interests of the biophysical movement of 1847', *J. Hist. Med.* xxi (1966).

Crew, Henry, 'Helmholtz on the doctrine of energy', *J. Opt. Soc. Amer.* 6 (1922) 312–326.

Crombie, A. C., 'Helmholtz', *Sci. Amer.* 198 (1958) 94–102.

Dingler, Hugo, 'H. Helmholtz und die Grundlagen der Geometrie', *Z. Phys.* 90 (1934) 348–354 and 94 674–676.

Dingler, Hugo, 'Über den Zirkel in der empirischen Begründung der Geometrie', *Kant-Studien* 30 (1925) 310–330.

Ebert, Hermann, *Hermann von Helmholtz* (Stuttgart, 1949) 199 pp.

Elkana, Yehuda, *The Discovery of the Conservation of Energy* (Cambridge, Mass. and London, 1974) 213 pp.

Elkana, Yehuda, 'Helmholtz' *Kraft*: An illustration of concepts in flux', *Hist. Studies Phys. Sci.* 2 (1970) 263–298.

Engelmann, Theodor W., *Gedächtnisrede auf Hermann von Helmholtz* (Leipzig, 1894) 34 pp.

Engels, Friedrich, *Dialectics of Nature*, tr. C. Dutt (many editions, e.g. Moscow and London, 1954) pp. 99–136 and *passim*.

Epstein, S., *Helmholtz als Mensch und Denker* (Stuttgart, 1896).

Erdmann, Benno, 'Die philosophische Grundlagen von Helmholtz' Wahrnehmungstheorie, kritisch erläutert', *Abhandlungen der Preussischen Akad. der Wiss., phil.-hist. Klasse* (Berlin, 1921) Nr. 1, pp. 1–45.

Frankfurt, U. I., 'The electrodynamics of Helmholtz and its evolution', *Voprosy Istorii Estestvoznaniia i Techniki* (1963) No. 14, pp. 49–55 (in Russian).

Goldschmidt, Ludwig, *Kant und Helmholtz – populärwissenschaftliche Studie* (Hamburg and Leipzig, 1898) 135 pp.

Grell, Heinrich, 'Über das Helmholtz-Liesche Raumproblem', in Klare *et al.* (1972) pp. 69–78.

Gross, Theodor, *Kritische Beiträge zur Energetik* (Berlin, 1902). Concerning Helmholtz, pp. 59–236.

Gross, Theodor, *Robert Mayer und Hermann von Helmholtz: Eine kritische Studie* (Berlin, 1898) 174 pp.

d'Haëne, Robert, 'La notion scientifique de l'energie, son origine et ses limites', *Rev. de Metaphys. et de morale* **72** (1967) 35–67.

Hall, G. Stanley, *Founders of Modern Psychology* (N.Y., 1912). Ch. on Helmholtz, pp. 247–308.

Hamm, Josef, *Das philosophische Weltbild von Helmholtz* (Bielefeld, 1937) 87 pp. Inaugural dissertation at Bonn.

Heimann, P., 'Helmholtz and Kant: The metaphysical foundations of *Über die Erhaltung der Kraft*', *Studies Hist. Phil. Sci.* **5** (1974) 205–238.

Helm, Georg, *Die Energetik nach ihrer geschichtlichen Entwickelung* (Leipzig, 1898). Part 1 Sect. 8, 'Helmholtz' standpunkt'; Sect. 9, 'Helmholtz' Anwendungen der Energetik'; Part 5 Sect. 7, 'Helmholtz' thermochemische Arbeiten'; Part 8, Sect. 1, 'Die Lagrange'-schen Differentialgleichungen'.

Herneck, Friedrich, 'Die Stellung von Hermann von Helmholtz in der Wissenschafts-geschichte', *Wiss. Z. Humboldt-Univ. Berlin, Math.-natur. Reihe* **22** (1973) pp. 349–355.

Hertz, Paul (See Schlick below).

Heyfelder, Victor, *Über den Begriff der Erfahrung bei Helmholtz* (Berlin, 1897) 81 pp. Inaugural dissertation at Berlin.

Höller, Hubert, *Der Gegenstand der Sinneswahrnehmung bei Aristoteles und Helmholtz: Zur Geschichte und Erläuterung des Wahrnehmungsproblems* (Giessen, 1925). Dissertation.

Hörz, Herbert, ch. on Helmholtz in *Von Liebig zu Laue. Ethos und Weltbild grosser deutscher Naturforscher und Ärzte* (2nd ed., Berlin, 1963) pp. 111–122.

Hörz, Herbert and Siegfried Wollgast, 'Einleitung' (79 pp.) and 'Anmerkungen' to their edition of Helmholtz' *Philosophische Vorträge und Aufsätze* (Berlin, 1971).

Jones, Ernest, *The Life and Work of Sigmund Freud*, vol. one *passim* (N.Y. and London, 1953). On Freud's early scientific education by the 'school of Helmholtz'. See also Bernfeld, above.

Kahl, H. Russell, *The Philosophical Work of Hermann von Helmholtz*. Ph.D. dissertation, Columbia University, 1951.

Karpinski, Louis C., 'Hermann von Helmholtz', *Sci. Monthly* **13** (1921) 23–31.

Keeton, M.T., 'Some ambiguities in the theory of the conservation of energy', *Phil. Sci.* **8** (1941) 304–319.

Kirchhoff, 'Helmholtz, A Memoir', *Ann. Report Smithson. Inst.* for 1889 (Washington, D.C., 1900) pp. 527–540.

Klare, H. *et al.*, *Gedanken von Helmholtz über schöpferische Impulse und über des Zusam-menwirken verschiedener Wissenschaftszweige, Sitzungsber. des Plenums u. d. Kl. d. Akad. d. Wiss. d. DDR* (1972, No. 1) 78 pp. Lectures on the 150th anniversary of the birth of Helmholtz; titles listed separately by authors as follows: Werner Albring, Friedhart Klix, Hans-Jürgen Treder, Heinz Stiller, Alfred Kosing, Robert Rompe, Karl-Friedrich Alexander, Heinrich Grell.

Klix, Friedhart, 'H. v. Helmholtz' Beitrag zur Theorie der Wahrnehmung', in Klare *et al.* (1972) pp. 25–36.

Koenigsberger, Leo, *Hermann von Helmholtz's Untersuchungen über die Grundlagen der Mathematik und Mechanik* (Leipzig, 1896) 58 pp. English tr. in *Ann. Reports Smithson. Inst.* for 1896 (Washington, D.C., 1898) pp. 93–124.

Koenigsberger, Leo, 'Über Helmholtz' Bruchstück eines Entwurf, betitelt "Naturforscher-rede"', *Sitz. Ber. Heidelberger Akad. Wiss., Math.-natur. Kl.* no. 1, 14 Abhandl. (1910).

König, Arthur Peter, ed., *Beiträge zur Psychologie und Physiologie der Sinnesorgane* (Hamburg, 1891) 388 pp. Essays to celebrate Helmholtz' 70th birthday; among them: Preyer, 'Ursprung des Zahlbegriffes aus dem Tonsinn'.

Lipps, 'Aesthetischen Faktoren der Raumanschauung'.

Kosing, Alfred, 'Zu den philosophischen Anschauungen von Helmholtz', in Klare *et al.* (1972) pp. 49–55.

Krause, Albrecht, *Kant und Helmholtz über den Ursprung und die Bedeutung der Raumanschauung und der geometrischen Axiome* (Lahr, 1878) 102 pp.

Krebs, G., *Die Erhaltung der Energie als Grundlage der neueren Physik* (Munich, 1877).

Kronecker, Hugo, *Hermann von Helmholtz* (Bern, 1894). Reprinted from the *Schweizerischen Rundschau* (Zürich, 1894) 31 pp.

Kuehtmann, Alfred, *Zur Geschichte der Terminismus: Occam, Condillac, Helmholtz, Mauthner* (Leipzig, 1911). *Abhandl. z. Phil. u. ihre Geschichte*, Heft 20, 127 pp.

Kuhn, Thomas, 'Energy conservation as an example of simultaneous discovery', in *Critical Problems in the History of Science*, ed. M. Clagett (Madison, 1969), pp. 321–356.

Kuznetsov, B.G. and U.I. Frankfurt, 'History of the law of conservation and transmutation of energy', *Trudy Inst. Istorii Estestvoznaniia i Techniki* **28** (1959) 339–376. Particularly concerned with Helmholtz and Robert Mayer.

Lamb, Horace, 'Reciprocal relations in dynamics', *Proc. London Math. Soc.* **19** (1888).

Laue, Max von, 'Über Hermann von Helmholtz' in *Festschrift zur 150-Jahr-Feier der Humboldt-Universität zu Berlin, 1810–1960*, ed. Friedrich Herneck, Vol. 1 (Berlin, 1960).

Lazarev, Peter P., *Helmholtz* (Moscow, 1959) 102 pp. (in Russian).

Lebedinskii, Andrei V., U.I. Frankfurt and A.M. Frenk, *Helmholtz (1812–1894)* (Moscow, 1966, in Russian) 314 pp. with a brief Afterword about the senior author, A.V. Lebedinskii (1902–1965).

Lenin, V.I., *Materialism and Empirio-Criticism*, tr. A. Fineberg (many editions, e.g. *Collected Works* **14**, Moscow, 1962), esp. Ch. 4 Sect. 6, 'The "theory of symbols" and the criticism of Helmholtz'.

Lenzen, Victor F., 'Helmholtz's Theory of Knowledge', *Studies and Essays in the History of Science and Learning Offered in Homage to George Sarton On the Occasion of his 60th Birthday*, ed. M.F. Montagu (N.Y., 1946) pp. 301–319.

Liebert, Arthur, 'Johannes Müller, der Physiologe, in seinem Verhältnis zur Philosophie und in seiner Bedeutung für dieselbe', *Kant-Studien* **20** (1915) 357–375.

Mandelbaum, Maurice, *History, Man and Reason: A Study in Nineteenth-Century Thought* (Baltimore, 1971). Chapter 14 'Ignoramus, Ignorabimus: The Positivist Strand', Sect. 1 'Helmholtz: science and epistemology'.

Maxwell, J. Clerk, 'Helmholtz' ['Scientific Worthies: x'], *Nature* **15** (1877) 389–391; and in *Sci. Papers*, vol. 2, ed. W.D. Niven (Cambridge, England, 1890; Dover reprint, N.Y., 1965) pp. 592–598.

Mendelsohn, Everett, 'Physical models and physiological concepts: explanation in nineteenth century biology', *Boston Studies in the Philosophy of Science*, volume 2 (Atlantic

Highlands, N.J., Humanities Press, 1965) pp. 127–150.

Menselsohn, Everett, 'The biological sciences in the nineteenth century: some problems and sources', *Hist. Sci.* **3** (1964).

Mendenhall, T.C., *Ann. Report Smithson. Inst.* for 1894 (Washington, D.C., 1895) pp. 787–793.

Merz, John Theodore, *A History of European Thought in the Nineteenth Century*, 4 vol. (Edinburgh and London, 1912, and 1930; Humanities reprint, N.Y., 1965). For Helmholtz, see volumes 3 and 4.

Meyerson, Emile, *Identity and Reality*, tr. Kate Loewenberg (London and N.Y., 1930; Dover reprint, N.Y., 1962).

Meyerson, Emile, *Du cheminement de la pensée*, 3 vol. (Paris, 1931).

M'Kendrick, John G., *Hermann Ludwig Ferdinand von Helmholtz* (London and N.Y., 1899) 299 pp. In the series "Masters of Medicine".

Moore, J., 'The conservation of energy not a fact, but a heresy of science', *Nature* (1872) p. 180.

Murphy, Gardner, *Historical Introduction to Modern Psychology*, rev. ed. (N.Y., 1949), Ch. 10 'German Physiological Psychology in the Age of Helmholtz'.

Naturwissenschaften **9** Heft 35, 31 August 1921. A special issue 'Dem Andenken an Helmholtz zur Jahrhundertfeier seines Geburtstages'. Among the essays:

Johannes von Kries, 'Helmholtz as Physiolog', pp. 673–693.

W. Wien, 'Helmholtz als Physiker', pp. 694–699.

W. Nernst, 'Die elektrochemischen Arbeiten von Helmholtz', pp. 699–702.

A. Riehl, 'Helmholtz als Erkenntnistheoretiker', pp. 702–708.

Neumann, C., *Die elektrischen Kräfte. Darlegung und Erweiterung der von A. Ampère, F. Neumann, W. Weber, G. Kirchhoff entwickelten mathematischen Theorien*, 2 vol. (Leipzig, 1873 and 1898), 272 and 462 pp. In vol. 2, see: 'Über die von Hermann von Helmholtz in seiner älteren und in seiner neueren Arbeiten angestellten Untersuchungen'.

Ostwald, Wilhelm, *Studien zur Biologie des Genies*, Vol. 1, *Grosse Männer* (Leipzig, 1910). On Helmholtz, pp. 256–310.

Pavlov, I.P., 'Natural Science and the Brain', ch. 10 of *Lectures on Conditioned Reflexes*, vol. one, tr. W. Horsley Gantt (N.Y., 1928) pp. 120–130.

Pielert, Walter, *Kritische Betrachtungen der erkenntnistheoretischen Ansichten von Helmholtz und Fick* (Halle, 1924).

Planck, Max, *Das Prinzip der Erhaltung der Energie*, 4th ed. (Berlin, 1921).

Polak, L.S., 'Hidden motion in Helmholtz' theory of heat', *Voprosy Istorii Estestvoznaniia i Tekhniki* (1957) No. 3, pp. 62–73 (in Russian).

Polak, L.S., and Yu. I. Solov'ev, 'On the history of physical chemistry. Helmholtz' studies in the field of physical chemistry', *Voprosy Istorii Estestvoznaniia i Tekhniki* (1959) No. 8, pp. 48–56 (in Russian).

Rau, Albrecht, *Empfinden und Denken. Eine physiologische Untersuchung über der Nature des menschlichen Verstandes* (Giessen, 1896). See ch. 10, 'Die Lehre von den Tonempfindungen nach H. v. Helmholtz', and ch. 11, 'Der Einfluss von Helmholtz' widersprüchiger Auffassung auf die Gestaltung anderer Theorien', pp. 213–282.

Reinecke, Wilhelm, 'Die Grundlagen der Geometrie', *Kant-Studien* **8** (1903) 345–395. A study of Newton, Kant, Gauss and Helmholtz.

Reiner, Julius, *Hermann von Helmholtz* (Leipzig, 1905) 203 pp.

Reiprich, K., *Die philosophisch-naturwissenschaftliche Arbeiten von Karl Marx und Friedrich Engels* (Berlin, 1969). For a discussion of the relation of Engels to Helmholtz.

Riehl, Alois, 'Helmholtz als Erkenntnistheoretiker', *Naturwiss.* **9** (1921) 702–708. An essay in the special issue devoted to Helmholtz on the centenary of his birth; see *Naturwissenschaften*.

Riehl, Alois, 'Helmholtz in seinem Verhältnis zu Kant', *Kant-Studien* **9** (1904) 261–285. Also published separately (Berlin, 1904).

Riese, W., and Arrington, G.E., 'The history of Johannes Mueller's doctrine of the specific energies of the senses: Original and later version', *Bull. Hist. Med.* **37** (1964).

Rompe, Robert, 'Einige Worte über Helmholtz und die Physikalisch-Technische Reichsanstalt', in Klare *et al.* (1972) pp. 57–64.

Rubner, M., (See Warburg below).

Rücker, A.W., 'Helmholtz', *Proc. Roy. Soc. London* **59** (1896) pp. xvii–xxx. Also in *Ann. Report Smithson. Inst.* for 1894 (Washington, D.C., 1895) pp. 709–718.

Sarton, G., 'The discovery of the law of conservation of energy', *Isis* **xiii** (1929) 18.

Schlick, M., and P. Hertz, 'Erläuterungen' to their edition of Helmholtz' *Schriften zur Erkenntnistheorie* (Berlin, 1921) 175 pp. English tr. by Malcolm Lowe in this volume.

Schlick, M., *Philosophical Papers 1910–1936*, tr. P. Heath, ed. H.L. Mulder and Barbara van der Velde-Schlick (Dordrecht and Boston, in press).

Schlick, Moritz, 'Helmholtz als Erkenntnistheoretiker', in E. Warburg *et al.* (1922), pp. 29–39.

Schwertschlager, Joseph, *Kant und Helmholtz erkenntnistheoretisch vergleichen* (Freiburg-im-Breisgau, 1883).

Stallo, J.B., *The Concepts and Theories of Modern Physics* (1st ed., N.Y., 1881; new ed. edited by P.W. Bridgman, Camb., Mass., 1960). See especially Ch. 12, 'Character and Origin of the Mechanical Theory (continued) – Its Exemplification of the Fourth Radical Error of Metaphysics'.

Stiller, Heinz, 'Zur Bedeutung der Arbeiten von H. v. Helmholtz für die geophysikalische Hydrodynamik und für die Physik des Erdinnern', in Klare *et al.* (1972) pp. 45–48.

Stumpf, Carl, 'Hermann von Helmholtz und die neuere Psychologie', *Arch. f. Gesch. d. Philosophie* **8** (1895) 303–314. English tr. in *Psych. Rev.* **2** (1895) 1–12.

Treder, Hans-Jürgen, 'Die Bedeutung von Helmholtz für die theoretische Physik', in Klare *et al.* (1972) pp. 37–44.

Turner, R. Steven, 'Helmholtz, Hermann von', *Dict. Sci. Biog.* **6** (1972) 241–253. An excellent summary account of the scientific achievements.

Wagner, Kurt, 'Zur richtigen Deutung der Helmholtzschen Zeichentheorie', *Deutsche Z. Phil.* **13** (1965) 162–172.

Warburg, E., M. Rubner and M. Schlick, *Helmholtz als Physiker, Physiologe und Philosoph* (Karlsruhe i. B., 1922). Three lectures on the 100th anniversary of the birth of Helmholtz, at the physical, physiological and philosophical societies at Berlin:
E. Warburg, 'Helmholtz als Physiker', pp. 1–14.
M. Rubner, 'Helmholtz als Physiologe', pp. 14–29.
M. Schlick, 'Helmholtz als Erkenntnistheoretiker', pp. 29–39.

Warren, Richard M. and Roslyn P. Warren, 'Introduction' and 'Comments' to their edition of *Helmholtz on Perception: Its Physiology and Development* (N.Y., 1968).

Wever, Ernst Glen and Merle Lawrence, *Physiological Acoustics* (Princeton, 1954).

Weyl, Hermann, 'Philosophie der mathematik und naturwissenschaft', *Handbuch der*

Philosophie (Berlin, 1926); rev. and augmented English ed., *Philosophy of Mathematics and Natural Science* (Princeton, 1949).

Wien, W., 'Hermann von Helmholtz zu seinem 25-jährigen Todestage', *Naturwissenschaften* **7** (1919) 645–648.

Winters, Stephen M., 'Helmholtz, Dynamics and the Unity of Science', *Hist. Studies Phys. Sci.* (forthcoming).

Wittich, D., 'Das Gesetz der sogenannten spezifischen Sinnesenergie und die beiden philosophischen Grundrichtungen', *Deutsche Z. Phil.* **12** (1964) 682–691.

Wollgast, Siegfried, 'H. v. Helmholtz in der Sicht M. Schlicks', *Rostocker Philos. Manuskripte*, Heft 8, Teil 2 (1970) 51–63.

Woodruff, Arthur E., 'Action at a distance in nineteenth century electrodynamics', *Isis* **53** (1962) 439–459.

Woodruff, Arthur E., 'The contributions of Hermann von Helmholtz to Electrodynamics', *Isis* **59** (1968) 300–311.

Wundt, W., *Erlebtes und Erkanntes* (Berlin, 1920). Pp. 155–169 discuss Wundt's own appraisal of his relations to Helmholtz.

Wussing, Hans, 'Der philosophische Kampf um den Energiesatz', *Naturwissenschaft, Tradition, Fortschritt*, Suppl. vol. to *Z. Gesch. Naturwiss., Technik u. Medicin* (Berlin, 1963).

Yourgrau, Wolfgang, and Stanley Mandelstam, *Variational Principles in Dynamics and Quantum Theory* (N.Y. and London, 2nd ed., 1960), especially Ch. 13, 'The Significance of Variational Principles in Natural Philosophy'.

PART D. INDEX TO CITED WORKS
(with English translation where known)

Avenarius, R., *Kritik der reinen Erfahrung* (Leipzig, 1907–08).
Avenarius, R., *Der menschliche Weltbegriff* (Leipzig, 1891).

Beltrami, E., *Saggio di Interpretazione della Geometrie Non-Euclidea* (Naples, 1848).
Beltrami, E., 'Teoria Fondamentale degli Spazii di Curvatura Costante', *Annali di Matematica*, 2nd series, II, pp. 232–255.
DuBois-Reymond, P., *Allgemeine Funktionslehre* (Tübingen, 1882).
Bolzano, B., *Wissenschaftslehre*, vol. 2 (Sulzbach, 1837) [Eng. tr. Rolf George, *Theory of Science*, abridged (Berkeley and Los Angeles, 1972)].
Bonola and Liebmann, *Die nichteuclidische Geometrie* (Leipzig, 1908) [Eng. tr., R. Bonola, *Non-Euclidean Geometry*, tr. H.S. Carslaw (Chicago, 1912; Dover reprint, N.Y., 1955)].

Carnap, R., *Logical Foundations of Probability* (2nd ed., Chicago, 1962).
Clifford, William Kingdon, *Mathematical Papers* (London, 1882).
Conrat, F., See Bibliography, Part C.
Couturat, L., *Les Principes des mathématiques* (Paris, 1905).

Dedekind, R., *Was sind und was sollen die Zahlen?* (3rd ed., Brunswick, 1911) [Eng. tr. W.W. Beman, *Essays on the Theory of Numbers* (Chicago, 1901; Dover reprint, N.Y., 1963)].
Delboeuf, J., *Prolégomènes philosophiques de la géométrie* (Paris, 1860).
Deutsch, R., *Philosophische Prinzipien der Mathematik* (Leipzig, 1908).
Dilthey, W., 'Beiträge zur Lösung der Frage vom Ursprung unseres Glaubens an die Realität der Aussenwelt und seinem Recht', *Sitzungsberichter der Berliner Akademie* (1890).
Dufour, *Bull. de la Soc. médicale de la Suisse Romande* (1876).

Ehrenfest-Afanasjewa, T., 'Physikalische Grössen', *Math.-Naturw. Blätter* **8** (1905).
Ehrenfest-Afanasjewa, T., 'Der Dimensionsbegriff und der analytische Bau physikalischer Gleichungen', *Math. Ann.* **77** (1916) 259–276.
Einstein, A., *Geometrie und Erfahrung* (Berlin, 1921) [Eng. tr. G.B. Jeffery and W. Perrett, 'Geometry and Experience' in A. Einstein, *Sidelights on Relativity* (London, 1922)].
Elsas, A., *Über die Psychophysik* (Marburg, 1886).
Eötvos, Pekar, and Fekete, [Hertz' reference to *Naturwissenschaften* **7** (1919) is erroneous; Hertz' source has not been located.]
Erdmann, Benno, See Bibliography, Part C.
Erdmann, Benno, *Die Axiome der Geometrie* (Leipzig, 1877) ch. III.

Fichte, J.G., *Grundlage der gesamten Wissenschaftslehre* (1794–5, 1802) [Eng. tr. P. Heath and J. Lachs, *Fichte: Science of Knowledge* (N.Y., 1970)].
Fichte, J.G., *Die Bestimmung des Menschen* (1800).
Frege, G., *Grundgesetze der Arithmetik*, 2 vol. (Jena, 1893 and 1903) [Eng. tr. with intro. M. Furth, *The Basic Laws of Arithmetic* (Berkeley and Los Angeles, 1964)].
Frege, G., *Die Grundlagen der Arithmetik* (Breslau, 1884) [Eng. tr. *The Foundations of Arithmetic* (Oxford, 2nd rev. ed., 1953; N.Y., 1960)].

Gauss, C. F., *Disquisitiones* (of 1828), in *Ges. Werke* (Göttingen, 1863–71), vol. 4.
Gentzen, G., 'The freedom from contradiction of pure number theory', *Math. Ann.* **112** (1936) 493–565.
Grassmann, H., *Die Ausdehnungslehre* (Leipzig, 1844; 2nd ed. 1878).
Grassmann, R., *Die Formenlehre oder Mathematik* (Stettin, 1872).

Hilbert, D., *Grundlagen der Geometrie*, 3rd ed. (Leipzig-Berlin, 1909) [Eng. tr. L. Unger, *Foundations of Geometry*, 2nd ed. (LaSalle, Ill., 1971) from the 10th German ed., ed. P. Bernays].
Hillebrand, Franz, *Ewald Hering: Ein Gedenkwort der Psychophysik* (Berlin, 1918).
Hume, D., *A Treatise of Human Nature* (London, 1739; Selby-Bigge ed., Oxford, 1888).

Jevons, Wm. Stanley, *The Principles of Science* (London, 1874).

Kant, I., *Kritik der reinen Vernunft* (1781) [Eng. tr. Norman Kemp Smith, *Critique of Pure Reason* (London, 1929)].
Kant, I., *Die metaphysischen Anfangsgründe der Naturwissenschaft* (1786) [Eng. tr. J. Ellington, *Metaphysical Foundations of Natural Science* (Indianapolis and N.Y., 1970)].
Kant, I., *Prolegomena* (1783) [Eng. tr. P. Carus, *Prolegomena* (Chicago, 1933)].
Killing, W. K. J., *Grundlagen der Geometrie* (Paderborn, 1893).
Kirchhoff, G., *Mechanik* (first part of *Vorlesungen über Mathematischen Physik*), 3rd ed. (Leipzig, 1883).
Klein, F., *Gesammelte Abhandlungen*, vol. 1.
Krause, A., See Bibliography, Part C.

Land, J. P., 'Kant's space and modern mathematics', *Mind* **2** (1877) 38–46.
Leibniz, G. W., *Opera*, ed. J. E. Erdmann (Berlin, 1840) Part I. [For Leibniz' discussion on this and related matters, see L. E. Loemker ed., *Philosophical Papers and Letters* (Chicago, 1956; Reidel reprint, Dordrecht, 1969)].
Lie, S., *Theorie der Transformationsgruppen* (Leipzig, 1873), vol. 3, pp. 437–523.
Lipschitz, R., 'Untersuchungen über die ganzen homogenen Funktionen von *n* Differentialen', *Borchardts Journal f. Mathematik* **70** (1869) 71; **72**, 1.
Lipschitz, R., 'Untersuchung eines Problems der Variationsrechnung', *ibid*, **74**.
Lobatchewsky, N. J., *Prinzipien der Geometrie* (Kazan, 1829–30) [Eng. tr. H. S. Carslaw, *Theory of Parallels* (Chicago, 1914), also as appendix to Dover reprint of Bonola, *Non-Euclidean Geometry*, above].
Locke, J., *An Essay Concerning Human Understanding* (London, 1690; Pringle-Pattison .ed., Oxford, 1924).

Mach, E., *Beiträge zur Analyse der Empfindungen* (Jena, 1886) [Eng. tr. C. M. Williams, *The Analysis of Sensations*, 5th ed. (London, 1911)].
Mach, E., *Die Geschichte und die Würzel des Satzes von der Erhaltung der Arbeit* (Prague, 1872) [Eng. tr. Philip E. B. Jourdain, *History and Root of the Principle of the Conservation of Energy* (Chicago, 1911)].
Mach, E., *Die Mechanik in ihrer Entwickelung historisch-kritisch dargestellt* (Leipzig, 1883) [Eng. tr. of 2nd German ed., Thomas J. McCormack, *The Science of Mechanics* (Chicago and London, 1893; 6th ed. with new intro. by Karl Menger, LaSalle, Ill., 1960)].
Mendelson, Elliott, *Introduction to Mathematical Logic* (Princeton, 1964).
Meumann, E., 'Frage der Sensibilität der inneren Organe', *Arch. f. die gesamte Psychologie* **9** (1907) 26–62.

Michelson, A. A., 'Détermination expérimental de la valeur du métre en longueurs d'Ondes lumineuses', *Travaux et mém. du bureau internationale des poids et mesures* **11** (Paris, 1894).

Mongré, Paul [pseudonym for Felix Hausdorff], *Das Chaos in kosmischer Auslese* (Leipzig, 1898).

Newton, I., *Opticks*, 4th ed. (London, 1730; Dover reprint, N.Y., 1952).

Newton, I., *Principia* (London, 1687) [Eng. tr. Andrew Motte, *Mathematical Principles* (1729); rev. tr. F. Cajori (Berkeley, 1946)].

Pasch, M., *Vorlesungen über neuere Geometrie* (Berlin, 1882; 2nd ed. 1926).

Pasch, M., *Einleitung in die Differential- und Integralrechnung* (Leipzig, 1882).

Peano, G., *Notations de logique mathématique* (Turin, 1894).

Poincaré, H., *Der Wert der Wissenschaft* [Eng. tr. G. B. Halsted, 'The Value of Science', in *The Foundations of Science* (N.Y., 1929)].

Poincaré, H., *Wissenschaft und Hypothese* [Eng. tr. G. B. Halsted, 'Science and Hypothesis', in *The Foundations of Science* (N.Y., 1929)].

Poincaré, H., *Dernières Pensées* (Paris, 1913) [Eng. tr. J. W. Bolduc, *Mathematics and Science: Last Essays* (N.Y., 1963)].

Poincaré, H., *Science et Methode* (Paris, 1908) [Eng. tr. G. B. Halsted, 'Science and Method', in *The Foundations of Science* (N.Y., 1929)].

Reichenbach, H., *Relativitätstheorie und Erkenntnis a priori* (Berlin, 1921) [Eng. tr. Maria Reichenbach, *The Theory of Relativity and A Priori Knowledge* (Berkeley and Los Angeles, 1965)].

Riehl, A., 'Helmholtz in seinem Verhältnis zu Kant', See Bibliography, Part C.

Riehl, A., *Der philosophische Kritizismus*, 2nd ed., vol. 1 (Leipzig, 1908).

Riemann, B., *Über die Hypothesen, welche der Geometrie zugrunde liegen*, in *Werke*, 2nd ed. (1892); new ed. by H. Weyl with notes (Berlin, 1919). [Eng. tr. 'On the Hypotheses Which Lie at the Foundation of Geometry', in W. K. Clifford, *Collected Mathematical Papers*; also in *Nature* **8** (1873) 14–36; and in D. E. Smith, *A Source Book in Mathematics*, pp. 411–425].

Riemann-Weber, *Partielle Differentialgleichungen der Physik*, 5th ed. (Brunswick, 1912).

Russell, B., *Principles of Mathematics* (Cambridge, Eng., 1903).

Schlick, M., *Raum und Zeit* (1920) [Eng. tr. H. L. Brose, *Space and Time in Contemporary Physics* (Oxford and N.Y., 1920)].

Schlick, M., *Allgemeine Erkenntnislehre*, 2nd ed. (Berlin, 1925) [Eng. tr. A. E. Blumberg, *General Theory of Knowledge* (N.Y. and Vienna, 1974)].

Schlick, M., 'Idealität des Raumes, Introjektion und psychophysisches Problem', *Vierteljahresschr. für wiss. Philosophie* **40** (1916).

Schlick, M., 'Kritizistische oder empiristische Deutung der neuen Physik?', *Kant-Studien* **26** (1924).

Schlick, M., 'Naturphilosophische Betrachtungen über das Kausalprinzip', *Die Naturwiss.* **8** (1920). [The above 3 papers will appear in Eng. tr. in M. Schlick, *Philosophical Papers 1910–1936* (*Vienna Circle Collection*: Reidel, Dordrecht and Boston, in preparation).]

Schopenhauer, A., *Über die vierfache Würzel des Satzes vom Zureichenden Grunde* (1813) [Eng. tr. E. F. J. Payne, *The Fourfold Root of the Principle of Sufficient Reason* (LaSalle, Ill., 1974)].

Schopenhauer, A., *Die Welt als Wille und Vorstellung* (1818) [Eng. tr. E.F.J. Payne, *The World as Will and Representation* (N.Y., 1966)].

Schröder, E., *Lehrbuch der Arithmetik und Algebra* (Leipzig, 1873).

Schrödinger, E., 'Grundlinien einer Theorie der Farbenmetrik im Tagessehen', *Ann. d. Phys.* **63** (1920) 397, 427, 481.

Sommerfeld, A., *Atombau und Spektrallinien* (Braunschweig, 1921) [Eng. tr. H.L. Brose, *Atomic Structure and Spectral Lines* (London, 1923, from the 3rd German ed.; London and N.Y., 1934, from the 5th German ed.)].

Stumpf, K., *Tonpsychologie*, 2 vol. (Leipzig, 1883 and 1890).

Tobias, Wilhelm, *Grenzen der Philosophie, constatirt gegen Riemann und Helmholtz, vertheidigt gegen von Hartmann und Lasker* (Berlin, 1875).

Vaihinger, H., *Die Philosophie des Als-Ob* (Berlin, 1911) [Eng. tr. C.K. Ogden, *The Philosophy of 'As If'*, 2nd ed. (London and N.Y., 1935)].

Weyl, H., *Raum, Zeit, Materie*, 4th ed. (Berlin, 1921) [Eng. tr. H.L. Brose, *Space, Time, Matter* (N.Y., 1922)].

Weyl, H., See also Bibliography, Part C.

Weyl, H., 'Reine Infinitesimalgeometrie', *Math. Zeits.* **2** (1918) 384–411.

Whitehead, A.N. and B. Russell, *Principia Mathematica*, vol. 2 (Cambridge, Eng., 1912).

Wundt, W., *Grundzüge der Physiologischen Psychologie*, 5th ed., vol. II (Leipzig, 1902–03).

Zermelo, Ernst, 'Über die Grundlagen der Arithmetik', *Atti del IV Congresso Internazionale dei Matematici, Rome 1908* (Rome, 1909), pp. 8–11.

INDEX OF NAMES

SYNTHESE LIBRARY

Monographs on Epistemology, Logic, Methodology,
Philosophy of Science, Sociology of Science and of Knowledge, and on the
Mathematical Methods of Social and Behavioral Sciences

Managing Editor:
JAAKKO HINTIKKA (Academy of Finland and Stanford University)

Editors:

ROBERT S. COHEN (Boston University)
DONALD DAVIDSON (University of Chicago)
GABRIËL NUCHELMANS (University of Leyden)
WESLEY C. SALMON (University of Arizona)

1. J. M. Bocheński, *A Precis of Mathematical Logic.* 1959, X + 100 pp.
2. P. L. Guiraud, *Problèmes et méthodes de la statistique linguistique.* 1960, VI + 146 pp.
3. Hans Freudenthal (ed.), *The Concept and the Role of the Model in Mathematics and Natural and Social Sciences, Proceedings of a Colloquium held at Utrecht, The Netherlands, January 1960.* 1961, VI + 194 pp.
4. Evert W. Beth, *Formal Methods. An Introduction to Symbolic Logic and the Study of Effective Operations in Arithmetic and Logic.* 1962, XIV + 170 pp.
5. B. H. Kazemier and D. Vuysje (eds.), *Logic and Language. Studies Dedicated to Professor Rudolf Carnap on the Occasion of His Seventieth Birthday.* 1962, VI + 256 pp.
6. Marx W. Wartofsky (ed.), *Proceedings of the Boston Colloquium for the Philosophy of Science, 1961-1962,* Boston Studies in the Philosophy of Science (ed. by Robert S. Cohen and Marx W. Wartofsky), Volume I. 1973, VIII + 212 pp.
7. A. A. Zinov'ev, *Philosophical Problems of Many-Valued Logic.* 1963, XIV + 155 pp.
8. Georges Gurvitch, *The Spectrum of Social Time.* 1964, XXVI + 152 pp.
9. Paul Lorenzen, *Formal Logic.* 1965, VIII + 123 pp.
10. Robert S. Cohen and Marx W. Wartofsky (eds.), *In Honor of Philipp Frank,* Boston Studies in the Philosophy of Science (ed. by Robert S. Cohen and Marx W. Wartofsky), Volume II. 1965, XXXIV + 475 pp.
11. Evert W. Beth, *Mathematical Thought. An Introduction to the Philosophy of Mathematics.* 1965, XII + 208 pp.
12. Evert W. Beth and Jean Piaget, *Mathematical Epistemology and Psychology.* 1966, XII + 326 pp.
13. Guido Küng, *Ontology and the Logistic Analysis of Language. An Enquiry into the Contemporary Views on Universals.* 1967, XI + 210 pp.
14. Robert S. Cohen and Marx W. Wartofsky (eds.), *Proceedings of the Boston Colloquium for the Philosophy of Science 1964-1966, in Memory of Norwood Russell Hanson,* Boston Studies in the Philosophy of Science (ed. by Robert S. Cohen and Marx W. Wartofsky), Volume III. 1967, XLIX + 489 pp.

15. C. D. Broad, *Induction, Probability, and Causation. Selected Papers.* 1968, XI + 296 pp.
16. Günther Patzig, *Aristotle's Theory of the Syllogism. A Logical-Philosophical Study of Book A of the Prior Analytics.* 1968, XVII + 215 pp.
17. Nicholas Rescher, *Topics in Philosophical Logic.* 1968, XIV + 347 pp.
18. Robert S. Cohen and Marx W. Wartofsky (eds.), *Proceedings of the Boston Colloquium for the Philosophy of Science 1966-1968,* Boston Studies in the Philosophy of Science (ed. by Robert S. Cohen and Marx W. Wartofsky), Volume IV. 1969, VIII + 537 pp.
19. Robert S. Cohen and Marx W. Wartofsky (eds.), *Proceedings of the Boston Colloquium for the Philosophy of Science 1966-1968,* Boston Studies in the Philosophy of Science (ed. by Robert S. Cohen and Marx W. Wartofsky), Volume V. 1969, VIII + 482 pp.
20. J.W. Davis, D. J. Hockney, and W. K. Wilson (eds.), *Philosophical Logic.* 1969, VIII + 277 pp.
21. D. Davidson and J. Hintikka (eds.), *Words and Objections: Essays on the Work of W.V. Quine.* 1969, VIII + 366 pp.
22. Patrick Suppes, *Studies in the Methodology and Foundations of Science. Selected Papers from 1911 to 1969.* 1969, XII + 473 pp.
23. Jaakko Hintikka, *Models for Modalities. Selected Essays.* 1969, IX + 220 pp.
24. Nicholas Rescher *et al.* (eds.), *Essays in Honor of Carl G. Hempel. A Tribute on the Occasion of His Sixty-Fifth Birthday.* 1969, VII + 272 pp.
25. P. V. Tavanec (ed.), *Problems of the Logic of Scientific Knowledge.* 1969, XII + 429 pp.
26. Marshall Swain (ed.), *Induction, Acceptance, and Rational Belief.* 1970, VII + 232 pp.
27. Robert S. Cohen and Raymond J. Seeger (eds.), *Ernst Mach: Physicist and Philosopher,* Boston Studies in the Philosophy of Science (ed. by Robert S. Cohen and Marx W. Wartofsky), Volume VI. 1970, VIII + 295 pp.
28. Jaakko Hintikka and Patrick Suppes, *Information and Inference.* 1970, X + 336 pp.
29. Karel Lambert, *Philosophical Problems in Logic. Some Recent Developments.* 1970, VII + 176 pp.
30. Rolf A. Eberle, *Nominalistic Systems.* 1970, IX + 217 pp.
31. Paul Weingartner and Gerhard Zecha (eds.), *Induction, Physics, and Ethics: Proceedings and Discussions of the 1968 Salzburg Colloquium in the Philosophy of Science.* 1970, X + 382 pp.
32. Evert W. Beth, *Aspects of Modern Logic.* 1970, XI + 176 pp.
33. Risto Hilpinen (ed.), *Deontic Logic: Introductory and Systematic Readings.* 1971, VII + 182 pp.
34. Jean-Louis Krivine, *Introduction to Axiomatic Set Theory.* 1971, VII + 98 pp.
35. Joseph D. Sneed, *The Logical Structure of Mathematical Physics.* 1971, XV + 311 pp.
36. Carl R. Kordig, *The Justification of Scientific Change.* 1971, XIV + 119 pp.
37. Milič Čapek, *Bergson and Modern Physics,* Boston Studies in the Philosophy of Science (ed. by Robert S. Cohen and Marx W. Wartofsky), Volume VII. 1971, XV + 414 pp.

38. Norwood Russell Hanson, *What I Do Not Believe, and Other Essays* (ed. by Stephen Toulmin and Harry Woolf), 1971, XII + 390 pp.
39. Roger C. Buck and Robert S. Cohen (eds.), *PSA 1970. In Memory of Rudolf Carnap*, Boston Studies in the Philosophy of Science (ed. by Robert S. Cohen and Marx W. Wartofsky), Volume VIII. 1971, LXVI + 615 pp. Also available as paperback.
40. Donald Davidson and Gilbert Harman (eds.), *Semantics of Natural Language*. 1972, X + 769 pp. Also available as paperback.
41. Yehoshua Bar-Hillel (ed.), *Pragmatics of Natural Languages*. 1971, VII + 231 pp.
42. Sören Stenlund, *Combinators, λ-Terms and Proof Theory*. 1972, 184 pp.
43. Martin Strauss, *Modern Physics and Its Philosophy. Selected Papers in the Logic, History, and Philosophy of Science*. 1972, X + 297 pp.
44. Mario Bunge, *Method, Model and Matter*. 1973, VII + 196 pp.
45. Mario Bunge, *Philosophy of Physics*. 1973, IX + 248 pp.
46. A. A. Zinov'ev, *Foundations of the Logical Theory of Scientific Knowledge (Complex Logic)*, Boston Studies in the Philosophy of Science (ed. by Robert S. Cohen and Marx W. Wartofsky), Volume IX. Revised and enlarged English edition with an appendix, by G. A. Smirnov, E. A. Sidorenka, A. M. Fedina, and L. A. Bobrova. 1973, XXII + 301 pp. Also available as paperback.
47. Ladislav Tondl, *Scientific Procedures*, Boston Studies in the Philosophy of Science (ed. by Robert S. Cohen and Marx W. Wartofsky), Volume X. 1973, XII + 268 pp. Also available as paperback.
48. Norwood Russell Hanson, *Constellations and Conjectures* (ed. by Willard C. Humphreys, Jr.). 1973, X + 282 pp.
49. K. J. J. Hintikka, J. M. E. Moravcsik, and P. Suppes (eds.), *Approaches to Natural Language. Proceedings of the 1970 Stanford Workshop on Grammar and Semantics*. 1973, VIII + 526 pp. Also available as paperback.
50. Mario Bunge (ed.), *Exact Philosophy — Problems, Tools, and Goals*. 1973, X + 214 pp.
51. Radu J. Bogdan and Ilkka Niiniluoto (eds.), *Logic, Language, and Probability. A Selection of Papers Contributed to Sections IV, VI, and XI of the Fourth International Congress for Logic, Methodology, and Philosophy of Science, Bucharest, September 1971*. 1973, X + 323 pp.
52. Glenn Pearce and Patrick Maynard (eds.), *Conceptual Chance*. 1973, XII + 282 pp.
53. Ilkka Niiniluoto and Raimo Tuomela, *Theoretical Concepts and Hypothetico-Inductive Inference*. 1973, VII + 264 pp.
54. Roland Fraïssé, *Course of Mathematical Logic — Volume 1: Relation and Logical Formula*. 1973, XVI + 186 pp. Also available as paperback.
55. Adolf Grünbaum, *Philosophical Problems of Space and Time*. Second, enlarged edition, Boston Studies in the Philosophy of Science (ed. by Robert S. Cohen and Marx W. Wartofsky), Volume XII. 1973, XXIII + 884 pp. Also available as paperback.
56. Patrick Suppes (ed.), *Space, Time, and Geometry*. 1973, XI + 424 pp.
57. Hans Kelsen, *Essays in Legal and Moral Philosophy*, selected and introduced by Ota Weinberger. 1973, XXVIII + 300 pp.
58. R. J. Seeger and Robert S. Cohen (eds.), *Philosophical Foundations of Science. Proceedings of an AAAS Program, 1969*, Boston Studies in the Philosophy of

Science (ed. by Robert S. Cohen and Marx W. Wartofsky), Volume XI. 1974, X + 545 pp. Also available as paperback.

59. Robert S. Cohen and Marx W. Wartofsky (eds.), *Logical and Epistemological Studies in Contemporary Physics*, Boston Studies in the Philosophy of Science (ed. by Robert S. Cohen and Marx W. Wartofsky), Volume XIII. 1973, VIII + 462 pp. Also available as paperback.

60. Robert S. Cohen and Marx W. Wartofsky (eds.), *Methodological and Historical Essays in the Natural and Social Sciences. Proceedings of the Boston Colloquium for the Philosophy of Science, 1969-1972*, Boston Studies in the Philosophy of Science (ed. by Robert S. Cohen and Marx W. Wartofsky), Volume XIV. 1974, VIII + 405 pp. Also available as paperback.

61. Robert S. Cohen, J. J. Stachel and Marx W. Wartofsky (eds.), *For Dirk Struik. Scientific, Historical and Political Essays in Honor of Dirk J. Struik*, Boston Studies in the Philosophy of Science (ed. by Robert S. Cohen and Marx W. Wartofsky), Volume XV. 1974, XXVII + 652 pp. Also available as paperback.

62. Kazimierz Ajdukiewicz, *Pragmatic Logic*, transl. from the Polish by Olgierd Wojtasiewicz. 1974, XV + 460 pp.

63. Sören Stenlund (ed.), *Logical Theory and Semantic Analysis. Essays Dedicated to Stig Kanger on His Fiftieth Birthday*. 1974, V + 217 pp.

64. Kenneth F. Schaffner and Robert S. Cohen (eds.), *Proceedings of the 1972 Biennial Meeting, Philosophy of Science Association*, Boston Studies in the Philosophy of Science (ed. by Robert S. Cohen and Marx W. Wartofsky), Volume XX. 1974, IX + 444 pp. Also available as paperback.

65. Henry E. Kyburg, Jr., *The Logical Foundations of Statistical Inference*. 1974, IX + 421 pp.

66. Marjorie Grene, *The Understanding of Nature: Essays in the Philosophy of Biology*, Boston Studies in the Philosophy of Science (ed. by Robert S. Cohen and Marx W. Wartofsky), Volume XXIII. 1974, XII + 360 pp. Also available as paperback.

67. Jan M. Broekman, *Structuralism: Moscow, Prague, Paris*. 1974, IX + 117 pp.

68. Norman Geschwind, *Selected Papers on Language and the Brain*, Boston Studies in the Philosophy of Science (ed. by Robert S. Cohen and Marx W. Wartofsky), Volume XVI. 1974, XII + 549 pp. Also available as paperback.

69. Roland Fraïssé, *Course of Mathematical Logic* – Volume 2: *Model Theory*. 1974, XIX + 192 pp.

70. Andrzej Grzegorczyk, *An Outline of Mathematical Logic. Fundamental Results and Notions Explained with All Details*. 1974, X + 596 pp.

71. Franz von Kutschera, *Philosophy of Language*. 1975, VII + 305 pp.

72. Juha Manninen and Raimo Tuomela (eds.), *Essays on Explanation and Understanding. Studies in the Foundations of Humanities and Social Sciences*. 1976, VII + 440 pp.

73. Jaakko Hintikka (ed.), *Rudolf Carnap, Logical Empiricist. Materials and Perspectives*. 1975, LXVIII + 400 pp.

74. Milič Čapek (ed.), *The Concepts of Space and Time. Their Structure and Their Development*, Boston Studies in the Philosophy of Science (ed. by Robert S. Cohen and Marx W. Wartofsky), Volume XXII. 1976, LVI + 570 pp. Also available as paperback.

75. Jaakko Hintikka and Unto Remes, *The Method of Analysis. Its Geometrical Origin and Its General Significance*, Boston Studies in the Philosophy of Science (ed. by Robert S. Cohen and Marx W. Wartofsky), Volume XXV. 1974, XVIII + 144 pp. Also available as paperback.

76. John Emery Murdoch and Edith Dudley Sylla, *The Cultural Context of Medieval Learning. Proceedings of the First International Colloquium on Philosophy, Science, and Theology in the Middle Ages – September 1973,* Boston Studies in the Philosophy of Science (ed. by Robert S. Cohen and Marx W. Wartofsky), Volume XXVI. 1975, X + 566 pp. Also available as paperback.

77. Stefan Amsterdamski, *Between Experience and Metaphysics. Philosophical Problems of the Evolution of Science,* Boston Studies in the Philosophy of Science (ed. by Robert S. Cohen and Marx W. Wartofsky), Volume XXXV. 1975, XVIII + 193 pp. Also available as paperback.

78. Patrick Suppes (ed.), *Logic and Probability in Quantum Mechanics.* 1976, XV + 541 pp.

79. Hermann von Helmholtz: *Epistemological Writings. The Paul Hertz/Moritz Schlick Centenary Edition of 1921 with Notes and Commentary by the Editors.* (Newly translated by Malcolm F. Lowe. Edited with an Introduction and Bibliography, by Robert S. Cohen and Yehuda Elkana), Boston Studies in the Philosophy of Science (ed. by Robert S. Cohen and Marx W. Wartofsky), Volume XXXVII. 1977, XXXVIII+204 pp. Also available as paperback.

80. Joseph Agassi, *Science in Flux,* Boston Studies in the Philosophy of Science (ed. by Robert S. Cohen and Marx W. Wartofsky), Volume XXVIII. 1975, XXVI + 553 pp. Also available as paperback.

81. Sandra G. Harding (ed.), *Can Theories Be Refuted? Essays on the Duhem-Quine Thesis.* 1976, XXI + 318 pp. Also available as paperback.

82. Stefan Nowak, *Methodology of Sociological Research: General Problems.* 1977, XVIII + 504 pp.

83. Jean Piaget, Jean-Blaise Grize, Alina Szeminska, and Vinh Bang, *Epistemology and Psychology of Functions,* Studies in Genetic Epistemology, Volume XXIII. 1977, XIV+205 pp.

84. Marjorie Grene and Everett Mendelsohn (eds.), *Topics in the Philosophy of Biology,* Boston Studies in the Philosophy of Science (ed. by Robert S. Cohen and Marx W. Wartofsky), Volume XXVII. 1976, XIII + 454 pp. Also available as paperback.

85. E. Fischbein, *The Intuitive Sources of Probabilistic Thinking in Children.* 1975, XIII + 204 pp.

86. Ernest W. Adams, *The Logic of Conditionals. An Application of Probability to Deductive Logic.* 1975, XIII + 156 pp.

87. Marian Przełęcki and Ryszard Wójcicki (eds.), *Twenty-Five Years of Logical Methodology in Poland.* 1977, VIII + 803 pp.

88. J. Topolski, *The Methodology of History.* 1976, X + 673 pp.

89. A. Kasher (ed.), *Language in Focus: Foundations, Methods and Systems. Essays Dedicated to Yehoshua Bar-Hillel,* Boston Studies in the Philosophy of Science (ed. by Robert S. Cohen and Marx W. Wartofsky), Volume XLIII. 1976, XXVIII + 679 pp. Also available as paperback.

90. Jaakko Hintikka, *The Intentions of Intentionality and Other New Models for Modalities.* 1975, XVIII + 262 pp. Also available as paperback.

91. Wolfgang Stegmüller, *Collected Papers on Epistemology, Philosophy of Science and History of Philosophy*, 2 Volumes, 1977, XXVII + 525 pp.
92. Dov M. Gabbay, *Investigations in Modal and Tense Logics with Applications to Problems in Philosophy and Linguistics*. 1976, XI + 306 pp.
93. Radu J. Bogdan, *Local Induction*. 1976, XIV + 340 pp.
94. Stefan Nowak, *Understanding and Prediction: Essays in the Methodology of Social and Behavioral Theories*. 1976, XIX + 482 pp.
95. Peter Mittelstaedt, *Philosophical Problems of Modern Physics*, Boston Studies in the Philosophy of Science (ed. by Robert S. Cohen and Marx W. Wartofsky), Volume XVIII. 1976, X + 211 pp. Also available as paperback.
96. Gerald Holton and William Blanpied (eds.), *Science and Its Public: The Changing Relationship*, Boston Studies in the Philosophy of Science (ed. by Robert S. Cohen and Marx W. Wartofsky), Volume XXXIII. 1976, XXV + 289 pp. Also available as paperback.
97. Myles Brand and Douglas Walton (eds.), *Action Theory. Proceedings of the Winnipeg Conference on Human Action, Held at Winnipeg, Manitoba, Canada, 9-11 May 1975*. 1976, VI + 345 pp.
98. Risto Hilpinen, *Knowledge and Rational Belief*. 1978 (forthcoming).
99. R. S. Cohen, P. K. Feyerabend, and M. W. Wartofsky (eds.), *Essays in Memory of Imre Lakatos*, Boston Studies in the Philosophy of Science (ed. by Robert S. Cohen and Marx W. Wartofsky), Volume XXXIX. 1976, XI + 762 pp. Also available as paperback.
100. R. S. Cohen and J. J. Stachel (eds.), *Selected Papers of Léon Rosenfeld*, Boston Studies in the Philosophy of Science (ed. by Robert S. Cohen and Marx W. Wartofsky), Volume XXI. 1977, XXX + 927 pp.
101. R. S. Cohen, C. A. Hooker, A. C. Michalos, and J. W. van Evra (eds.), *PSA 1974: Proceedings of the 1974 Biennial Meeting of the Philosophy of Science Association*, Boston Studies in the Philosophy of Science (ed. by Robert S. Cohen and Marx W. Wartofsky), Volume XXXII. 1976, XIII + 734 pp. Also available as paperback.
102. Yehuda Fried and Joseph Agassi, *Paranoia: A Study in Diagnosis*, Boston Studies in the Philosophy of Science (ed. by Robert S. Cohen and Marx W. Wartofsky), Volume L. 1976, XV + 212 pp. Also available as paperback.
103. Marian Przełęcki, Klemens Szaniawski, and Ryszard Wójcicki (eds.), *Formal Methods in the Methodology of Empirical Sciences*. 1976, 455 pp.
104. John M. Vickers, *Belief and Probability*. 1976, VIII + 202 pp.
105. Kurt H. Wolff, *Surrender and Catch: Experience and Inquiry Today*, Boston Studies in the Philosophy of Science (ed. by Robert S. Cohen and Marx W. Wartofsky), Volume LI. 1976, XII + 410 pp. Also available as paperback.
106. Karel Kosík, *Dialectics of the Concrete*, Boston Studies in the Philosophy of Science (ed. by Robert S. Cohen and Marx W. Wartofsky), Volume LII. 1976, VIII + 158 pp. Also available as paperback.
107. Nelson Goodman, *The Structure of Appearance*, Boston Studies in the Philosophy of Science (ed. by Robert S. Cohen and Marx W. Wartofsky), Volume LIII. 1977, L + 285 pp.
108. Jerzy Giedymin (ed.), *Kazimierz Ajdukiewicz: The Scientific World-Perspective and Other Essays, 1931 - 1963*. 1978, LIII + 378 pp.

109. Robert L. Causey, *Unity of Science*. 1977, VIII+185 pp.
110. Richard E. Grandy, *Advanced Logic for Applications*. 1977, XIV + 168 pp.
111. Robert P. McArthur, *Tense Logic*. 1976, VII + 84 pp.
112. Lars Lindahl, *Position and Change: A Study in Law and Logic*. 1977, IX + 299 pp.
113. Raimo Tuomela, *Dispositions*. 1978, X + 450 pp.
114. Herbert A. Simon, *Models of Discovery and Other Topics in the Methods of Science*, Boston Studies in the Philosophy of Science (ed. by Robert S. Cohen and Marx W. Wartofsky), Volume LIV. 1977, XX + 456 pp. Also available as paperback.
115. Roger D. Rosenkrantz, *Inference, Method and Decision*. 1977, XVI + 262 pp. Also available as paperback.
116. Raimo Tuomela, *Human Action and Its Explanation. A Study on the Philosophical Foundations of Psychology*. 1977, XII + 426 pp.
117. Morris Lazerowitz, *The Language of Philosophy, Freud and Wittgenstein*, Boston Studies in the Philosophy of Science (ed. by Robert S. Cohen and Marx W. Wartofsky), Volume LV. 1977, XVI + 209 pp.
118. Tran Duc Thao, *Origins of Language and Consciousness*, Boston Studies in the Philosophy of Science (ed. by Robert S. Cohen and Marx. W. Wartofsky), Volume LVI. 1977 (forthcoming).
119. Jerzy Pelc, *Semiotics in Poland, 1894 – 1969*. 1977, XXVI + 504 pp.
120. Ingmar Pörn, *Action Theory and Social Science. Some Formal Models*. 1977, X + 129 pp.
121. Joseph Margolis, *Persons and Minds, The Prospects of Nonreductive Materialism*, Boston Studies in the Philosophy of Science (ed. by Robert S. Cohen and Marx W. Wartofsky), Volume LVII. 1977, XIV + 282 pp. Also available as paperback.

SYNTHESE HISTORICAL LIBRARY

Texts and Studies
in the History of Logic and Philosophy

Editors:

N. KRETZMANN (Cornell University)
G. NUCHELMANS (University of Leyden)
L. M. DE RIJK (University of Leyden)

1. M. T. Beonio-Brocchieri Fumagalli, *The Logic of Abelard.* Translated from the Italian. 1969, IX + 101 pp.
2. Gottfried Wilhelm Leibniz, *Philosophical Papers and Letters.* A selection translated and edited, with an introduction, by Leroy E. Loemker. 1969, XII + 736 pp.
3. Ernst Mally, *Logische Schriften,* ed. by Karl Wolf and Paul Weingartner. 1971, X + 340 pp.
4. Lewis White Beck (ed.), *Proceedings of the Third International Kant Congress.* 1972, XI + 718 pp.
5. Bernard Bolzano, *Theory of Science,* ed. by Jan Berg. 1973, XV + 398 pp.
6. J. M. E. Moravcsik (ed.), *Patterns in Plato's Thought. Papers Arising Out of the 1971 West Coast Greek Philosophy Conference.* 1973, VIII + 212 pp.
7. Nabil Shehaby, *The Propositional Logic of Avicenna: A Translation from al-Shifā: al-Qiyās,* with Introduction, Commentary and Glossary. 1973, XIII + 296 pp.
8. Desmond Paul Henry, *Commentary on De Grammatico: The Historical-Logical Dimensions of a Dialogue of St. Anselm's.* 1974, IX + 345 pp.
9. John Corcoran, *Ancient Logic and Its Modern Interpretations.* 1974, X + 208 pp.
10. E. M. Barth, *The Logic of the Articles in Traditional Philosophy.* 1974, XXVII + 533 pp.
11. Jaakko Hintikka, *Knowledge and the Known. Historical Perspectives in Epistemology.* 1974, XII + 243 pp.
12. E. J. Ashworth, *Language and Logic in the Post-Medieval Period.* 1974, XIII + 304 pp.
13. Aristotle, *The Nicomachean Ethics.* Translated with Commentaries and Glossary by Hypocrates G. Apostle. 1975, XXI + 372 pp.
14. R. M. Dancy, *Sense and Contradiction: A Study in Aristotle.* 1975, XII + 184 pp.
15. Wilbur Richard Knorr, *The Evolution of the Euclidean Elements. A Study of the Theory of Incommensurable Magnitudes and Its Significance for Early Greek Geometry.* 1975, IX + 374 pp.
16. Augustine, *De Dialectica.* Translated with Introduction and Notes by B. Darrell Jackson. 1975, XI + 151 pp.